ANIMALS IN DISASTERS

ANIMALS IN DISASTERS

DICK GREEN

ELSEVIER

Butterworth-Heinemann
An imprint of Elsevier

Butterworth-Heinemann is an imprint of Elsevier
The Boulevard, Langford Lane, Kidlington, Oxford OX5 1GB, United Kingdom
50 Hampshire Street, 5th Floor, Cambridge, MA 02139, United States

Notices
Knowledge and best practice in this field are constantly changing. As new research and experience broaden our understanding, changes in research methods, professional practices, or medical treatment may become necessary.

Practitioners and researchers must always rely on their own experience and knowledge in evaluating and using any information, methods, compounds, or experiments described herein. In using such information or methods they should be mindful of their own safety and the safety of others, including parties for whom they have a professional responsibility.

To the fullest extent of the law, neither the Publisher nor the authors, contributors, or editors, assume any liability for any injury and/or damage to persons or property as a matter of products liability, negligence or otherwise, or from any use or operation of any methods, products, instructions, or ideas contained in the material herein.

Library of Congress Cataloging-in-Publication Data
A catalog record for this book is available from the Library of Congress

British Library Cataloguing-in-Publication Data
A catalogue record for this book is available from the British Library

ISBN: 978-0-12-813924-0

For information on all Butterworth-Heinemann publications visit our website at https://www.elsevier.com/books-and-journals

www.elsevier.com • www.bookaid.org

Publisher: Brian Romer
Acquisition Editor: Brian Romer
Editorial Project Manager: Hilary Carr
Production Project Manager: Anitha Sivaraj
Cover Designer: Vicky Pearson

Typeset by SPi Global, India

Contents

Preface

My first international response was to the Venezuela mudslides in December of 1999 often referred to as the Vargas Tragedy. Heavy torrential rains (nearly 3 ft) and flash floods caused a wall of earth and debris to come off the mountains killing tens of thousands of people, destroying thousands of homes and leaving an entire community under 9 ft of mud. I was managing the disaster response program for the American Humane Association at the time, and I received a call shortly after the event from a donor that had recently moved from Venezuela and was pleading that we assist in the animal rescue efforts. She was incredibly well connected in the country, and within 24 h, I was flying to Miami to catch a C-130 to Caracas. Upon landing, I met up with my counterparts at the World Society for the Protection of Animals (now World Animal Protection). There was a great need for urgency as the dogs that survived (many had sensed the earth movement and ran toward the ocean) were now looking for food and the soldiers were shooting them to keep them away from the human carcasses. So we engineered a plan, put together some teams, begged for boat and air support, and immediately began pulling scores of animals out the villages.

My Spanish is limited to certain beverages and facilities so communicating with the teams required patience and a really good interpreter. We had limited boats and only a promise of a Chinook helicopter, so we needed to be careful how we decided which animals would come out of the village each day. If they looked like they could be easily adopted, then they were first on the boat, and if they were fractious, they were left for a later day. On the other end, we were up against the challenge of what to do with all of the animals coming in. There was no shelter or even a rescue group that had the capacity or the capabilities to deal with 100+ dogs daily. So we identified a couple of tennis courts, and as we brought the dogs in, they were reunited with their friends while the local groups did an amazing job of finding new homes.

The village leadership agreed to a very small window for us to do our work, so we were working around the clock for five straight days. As our last day approached, we finally snagged the use of the Chinook that would allow us to get into the most remote village and pull out 50 of the neediest dogs. We crammed 25 cages and six responders into the helicopter, and they dropped us off early in the morning and promised to be back at 4 p.m. that afternoon. The team went to work searching for the dogs that were critical and adoptable. Only 50 would be coming out, and my job was to make sure that the amazing, compassionate responders did not exceed that number. As 3 p.m. rolled around, we brought the dogs down to the landing zone, and the vet provided a mild sedative knowing that there would be some anxiety as we crammed two to a kennel and then threw them into a helicopter. Most of these animals had never been in a house let alone a kennel inside a helo! I caught a few pups hidden in coats and backpacks, but all in all, we were ready at 4 p.m. for the trip back. We were still ready at 5 p.m. and at 6 p.m. when we finally got word that they would not be able to make it out to us until the following day. There were 50 dogs and 6 responders with little food and water and nothing to bed down in. In my go-bag, I always carry 100′ of rope so I strung it out on the beach, found two trees to anchor off, and tied an in-line figure 8 every couple of feet to stake the dogs out. No one slept including the dogs, but we were able to keep them safe until the morning when the bird returned. Obviously, our efforts were well received by the community and the nation, and as I was sitting in the airport to come home on Christmas Eve, a young Venezuelan girl brought me a Christmas card and a small gift. I knew I was in the right line of work.

Seems like every time that I present to a group about a disaster response like this, I hear the comment, "you should write a book." After 20 years and over 150 national and international responses, I have had the amazing opportunity to experience the worst of Mother Nature and the best of human and animal nature! So many examples of where things went well and an equal number of times that I wished I had done things differently. Emergency management refers to that as best practices and lessons learned. Throughout this text, the reader will be introduced to key components of animal rescue and incident management, and then, when appropriate, I will share a story of how that concept was applied and how well it worked for us.

Disasters are not going to go away—in fact, with changes in our climate, there is a good likelihood that we will see a number of new trends in disasters affecting the United States such as larger storms causing greater damage.[1] And the types of disasters we are seeing may indeed change with a greater number of flood and fire events, an increased number of winter storms, and a more active tornado season.[2]

Animal Search and Rescue (ASAR) is the newest member of the emergency management family. A number of key incidents occurred over the last 25 years that propelled the movement forward starting with Hurricane Andrew in 1992 and gaining full recognition with Hurricane Katrina in 2005 and the hurricane season of 2017 (Harvey, Irma, and Maria). Unlike our counterparts on the human side, we still have lots to learn about procedures, protocols, and best practices, but we have made great strides in a very short time.

In the United States, we do not have the consistency in the structure of Animal Search and Rescue (ASAR) teams like we do on the human side though we are making progress with recent strides in resource typing[3] and certification levels.[4] We also have ASAR teams that specialize in small or large animal and in water and land search and even teams that train specifically for an event such as fire or tornado response. Basic and technical guides presented in this text will need to be practiced and possibly adapted to fit your team's structure, capabilities, and response situations.

[1] Intergovernmental Panel on Climate Change www.ipcc.ch.

[2] *Science* 17 Oct 2014:Vol. 346, Issue 6207, pp. 349–352. https://doi.org/10.1126/science.1257460.

[3] SAADRA/NARSC Animal Resource Typing. NARSC.net.

[4] NASAAEP.org.

Definitions and Essential Acronyms

American Veterinary Medical Association (AVMA) The AVMA represents more than 91,000 members, and their mission is to protect, promote, and advance the needs of all veterinarians and those they serve.

Authority Having Jurisdiction (AHJ) The agency that has been assigned statutory authority; responsible for enforcing code and standards.

Cohabitated sheltering (CHS) People and their pets are physically together within a shelter.

Colocated sheltering (CLS) Colocated human/pet sheltering involves separate facilities for the humans and pets but is typically in close proximity so the humans can routinely visit and care for their pets.

Community/County Animal Response Team (CART) A network of animal and animal-agricultural resources under the auspices of the AHJ that are responsible for assisting a community in the planning, preparedness, response, and recovery efforts related to emergency incidents within a community.

Community Emergency Response Team (CERT) A program established by FEMA to educate individuals about disaster preparedness for hazards that may impact their area and train them in basic disaster response skills.

Comprehensive Emergency Management Plan (CEMP) Provides a policy-level framework to support emergency response activities within a jurisdiction.

Credentialing Identity and attributes of individuals or members of teams are validated against an established set of minimum criteria and qualifications for specific job titles.

Department of Homeland Security (DHS) A cabinet department of the US federal government with responsibilities in public security including antiterrorism, border security, immigration and customs, cybersecurity, and disaster prevention and management.

Disaster Any natural, technological, or civil event that causes injuries, deaths, or property damage of sufficient severity or magnitude to disrupt the essential functions and services (i.e., water supply, electric power, sanitation systems, roads, communication, and health services) of a community.

Emergency An event that causes injury or property damage beyond the capability of the victim(s) to handle without outside assistance.

Emergency management The process of preparing for mitigating, responding to, and recovering from an emergency or disaster.

Emergency operations center (EOC) A central command and control facility responsible for carrying out the principles of emergency preparedness and emergency management for a community during an emergency.

Emergency Support Function (ESF) An aspect of a disaster or emergency response assigned to a particular agency for management.

Federal Emergency Management Agency (FEMA) An agency of the US Department of Homeland Security, whose primary purpose is to coordinate the response to a disaster that overwhelms the resources of local and state authorities.

Hazard Something that can cause harm, for example, explosions, wind events, electricity, and chemicals.

Hazardous materials Any substance that has the potential to cause damage to the environment or the population if released. The substances are usually identified as being either flammable or combustible, explosive, toxic, noxious, corrosive, oxidizable, irritants, or radioactive.

Incident Command System (ICS) System used to organize and coordinate the field response to an emergency or disaster.

Mitigation The process of planning and preparation for the purpose of preventing the occurrence of a disaster or minimizing the severity of its impact.

National Alliance of State Animal and Agricultural Emergency Programs (NASAAEP) A collaborative alliance of state programs charged with planning, preparing for, and responding to disasters involving animals.

NASAAEP Best Practice Working Group (BPWG) NASAAEP, with funding from USDA Animal Care, convened a series of working groups (2010–14) that were charged with identifying best practices for various aspects of the emergency response for animals.

National Animal Rescue and Sheltering Coalition (NARSC) A coalition of national nongovernmental agencies that works collaboratively and cooperatively to assist communities and their animals in their preparations for and response to incidents that place animals in crisis.

National Capabilities for Animal Response in Emergencies (NCARE) An ASPCA study using a cross-sectional descriptive survey of officials who oversee emergency preparedness in US states and counties to determine state- and national-level animal response capabilities.

National Incident Management System (NIMS) NIMS provides a consistent nationwide template to enable all government, private-sector, and nongovernmental organizations to work together during domestic incidents.

National Integration Center (NIC) Responsible for the implementation and compliance of the National Incident Management System (NIMS). The NIC is designated to serve as an asset for government agencies, the private sector, and NGOs that are implementing and complying to NIMS guidelines.

National Response Framework (NRF) A guide to how the nation responds to all types of disasters and describes the principles, roles and responsibilities, and coordinating structures for delivering the core capabilities required to respond to an incident.

National Veterinary Response Team (NVRT) A component of the National Disaster Medical System under the US Department of Health and Human Services to provide veterinary medical care to ill or injured pets, service animals, working animals, laboratory animal, and livestock for augmenting the states' capabilities after a disaster.

Nongovernmental organizations (NGOs) A not-for-profit organization that is independent from county, state, and federal governmental organizations. They are usually funded by donations and oftentimes largely dependent on volunteers.

---

[1] https://www.fema.gov/media-library/assets/documents/15271.

Pet Evacuation Transportation Standards (PETS) Act Signed into law in October of 2006 and amended the Stafford Disaster Relief and Emergency Assistance Act[1] to, "ensure that State and local emergency preparedness operational plans address the needs of individuals with household pets and service animals following a major disaster or emergency."

Post-Katrina Emergency Management Reform Act (PKEMRA) Passed in 2006, strengthened the Federal Emergency Management Agency's (FEMA) preparedness and response capabilities, and identified new responsibilities for Department of Homeland Security (DHS) and FEMA in coordinating implementation of the PETS Act.

Preparedness Activities that enhance the abilities of individuals, communities, and businesses to better respond to a disaster.

Primary agency The agency or organization assigned primary responsibility to manage and coordinate a specific ESF. Primary agencies are responsible for overall planning and coordination with their support agencies and other ESFs.

Recovery Activities associated with the orderly restoration and rehabilitation of persons and property affected by disasters.

Resource typing Resource typing is defining and categorizing, by capability, the resources requested, deployed, and used in incidents.

Response Activities during and after a disaster that use all the systems, plans, and resources necessary to adequately reserve the health, safety, and welfare of victims and property affected by the disaster, with emphasis on meeting emergency needs and restoring essential community services.

Risk The chance, high or low, that any *hazard* will actually cause someone harm.

Southern Agricultural and Animal Disaster Response Alliance (SAADRA) An interactive collaboration of southern states at risk from similar natural, intentional, technological, and disease disasters affecting agriculture and animals.

State Animal Response Team (SART) Interagency state organizations or nongovernmental entities recognized by emergency management to prepare, plan, respond, and recover from animal emergencies in the United States.

Support agency Organization or agency designated to assist a primary agency with available resources, capabilities, or expertise to accomplish the mission of the ESF response and recovery operations under the coordination of the primary agency.

Technological hazard A range of hazards emanating from the manufacture, transportation, and use of hazardous materials, such as radioactives, chemical explosives, flammables, pesticides, herbicides, and disease agents; oil spills on land, coastal waters on inland water systems, and debris from space.

Weather *advisory* A regularly scheduled public news release issued by the national weather service providing details on a continuing weather event. Details include location, intensity, direction, and speed of movement of the event.

Weather *warning* A public news release issued by the national weather service indicating that a severe weather event is taking place or is imminent (within 24h or less) in a specified area. It is of utmost importance that all precautionary measures and actions be taken immediately for the protection of life and property.

Weather *watch* A public news release issued by the national weather service advising that conditions are present indicating possible development of a severe weather condition within a specified area. Preliminary disaster preparations should begin immediately, and TV, radio, and/or weather alert radio should be monitored for additional information and updates.

Wildland fires Any instance of uncontrolled burning in grasslands, brush, or wood lands. Wild fires can destroy property and valuable natural resources and may threaten the lives of people and animals.

CHAPTER

1

Introduction

Animals are a large part of our lives in the United States—in many households, they are considered one of the family. In the recent Black Forest Fire outside Colorado Springs (June, 2013), CNN interviewed a man that went back into the evacuation zone to rescue his dogs. When asked why he would put his own life in danger, he quickly responded that they were part of the family.

The American Veterinary Medical Association (AVMA) defines the human-animal bond as "a mutually beneficial and dynamic relationship between people and other animals that is influenced by behaviors that are essential to the health and well-being of both. This includes, but is not limited to, emotional, psychological, and physical interactions of people, other animals, and the environment." In their article entitled, "Placing the Human-Animal Bond in Context in the Face of Disasters" (May 4, 2006), they note that due to a lack of more traditional support systems in modern society, for many people, companion animals are the sole source of emotional and social support, providing significant psychological and physical health benefits, especially to children, the elderly, the disabled, the mentally and physically ill, and the incarcerated. Given this bond, they believe that "when disasters strike, saving animals means saving people."

The important role that animals play in our lives is not a new phenomenon, and there are examples of rescuing animals in distress dating back generations, but "animal rescue" as we know it today is a relatively new member to the disaster response field.

For the purpose of this text, a *disaster* is any event that causes injuries, deaths, or property damage of sufficient severity or magnitude to disrupt the essential functions and services of a community. A disaster overwhelms a community to a point where they are not able to meet all of the requests for assistance and outside help is needed. *Animal rescue* is defined as a component of animal welfare with a specific purpose of removing an animal from a situation that is or may become harmful. That might make Noah one of the first true animal rescuers (Fig. 1.1)! Unlike foster or adoption programs or traditional shelter work, animal rescue infers a *sense of emergency and a need for technical training and/or expertise.*

Animal welfare as we know it today got its start in 1822 when Irish politician, Richard Martin, was able to steer a bill through the English parliament offering protection of cruelty to cattle, horses, and sheep. Two years later, the first animal welfare organization was founded—the Society for the Prevention of Cruelty to Animals (SPCA). In 1840, Queen Victoria gave her blessing, and the SPCA became the Royal Society for the Prevention of Cruelty to Animals. In 1866, Henry Bergh brought the SPCA model to America, and the American Society for the Prevention of Cruelty to Animals (ASPCA) was founded.

In 1916, the American Humane Association (AHA) started its Red Star program with the specific purpose of helping military animals. Humane Society of the United States (HSUS) was founded in 1954 followed by International Fund for Animal Welfare (IFAW) in 1969, National Animal Control Association in 1978, United Animal Nations (now RedRover) in 1979, People for the Ethical Treatment of Animals (PETA) in 1980, Best Friends Humane Society in 1984, and Code 3 Associates in 1985.

It took a number of years and major disasters before the animal welfare groups developed specific animal rescue programs. A number of the groups responded to disasters, but there was very little attention to providing specific equipment, training, and financial support until the early 1990s following Hurricane Andrew (1992). AHA, HSUS, and United Animal Nations (UAN) were very active in the mid-1990s to late 1990s as they began to develop animal disaster teams, but there was very little effective communication, collaboration, or cooperation between the groups. It was an extremely competitive environment as the nongovernmental organizations (NGOs) vied for the lead role

Animals in Disasters
https://doi.org/10.1016/B978-0-12-813924-0.00001-3

FIG. 1.1   *Noah's Ark* by Edward Hicks, 1846.

and name recognition. In those days, since very few communities were active in animal rescue, whoever got to the scene first declared themselves as the lead agency and assumed "incident command." Self-deployment occurred too frequently; teams were not adequately trained in incident management or incident command; and yet, thousands of animals were rescued!

Not only was there poor communication on animal issues between the NGOs, but also a similar situation occurred between the NGOs and local government and the state, between the states and the Feds, and between the Feds and the NGOs. Two hurricanes changed that. In a span of 6 years, hurricanes Floyd and Katrina significantly changed the way that animal issues were addressed at the local, state, and federal levels and greatly impacted national animal response capabilities (Fig. 1.2).

Hurricane Floyd (1999) had a devastating effect on the state of North Carolina. According to the NC Department of Public Safety, approximately 2.8 million poultry, 2000 cattle, 250 horses, and more than 30,500 hogs drowned. It was the first time that we saw every major animal welfare group converge to provide support and, in many cases, the first time any of them had been asked to support livestock rescue efforts. Prior to this, the majority of the rescue groups worked

FIG. 1.2   Turkey Farm following Hurricane Floyd, 1999 (FEMA).

FIG. 1.3 Hurricane Katrina water rescue, 2005. *Courtesy of American Humane. All rights reserved.*

exclusively with dogs and cats and as a group had very little experience in handling livestock. Interestingly, even though there was very little organization, standardization of procedures, or interoperability (let alone a desire to communicate), thousands of animals were rescued from the floodwaters.

Following landfall of Hurricane Katrina (2005), things did not start out well in terms of the animal response as affected states (primarily Mississippi and Louisiana) were dealing with incredible human losses and animals were not a top priority. That soon changed as media captured animals on rooftops and pets being separated from their owners as they entered the evacuation bus. Katrina was the largest disaster in our history in terms of pets impacted, and after the first week, state, federal, and national animal resources were being dispatched to LA and MS. For the first time, NGOs were forced to work together, and state and federal resources were forced to work in the same camp as the NGOs. Granted, it was far from perfect, but it clearly changed the way that we respond to disasters today (Fig. 1.3).

Katrina created a situation that required the groups to play well together and for the states to recognize their role in coordinating those efforts and for enfolding them into human rescue operations. As a result, Katrina resulted in significant changes in the way that animals were prioritized and issues addressed. Without a doubt, countless animals were lost in Katrina and thousands of animals separated from their families, but because of the lessons learned, much fewer animals will be lost in future disasters. Since 2005, the United States has experienced a number of major disasters including hurricanes, devastating fires, deadly tornadoes, and historic floods. And without exception, the animal response has been better coordinated; communication has been more effective; and the collaboration between NGOs and state and federal partners has been vastly improved.

Two alliances were formed following Hurricane Katrina: the National Animal Rescue and Sheltering Coalition (NARSC) and the National Alliance of State Animal and Agricultural Emergency Programs (NASAAEP). In the years since Hurricane Katrina, NARSC (www.thenarsc.org) has developed and grown into a strong coalition whose mission is to identify, prioritize, and find collaborative solutions to major human-animal emergency issues. This coalition of 12 national organizations represents millions of animal welfare, animal care, and animal control professionals, volunteers, and pet owners. Participants in the coalition include some of the most experienced, qualified animal rescue and sheltering management professionals in the country.

The primary purpose of NASAAEP is to foster a national network of stakeholders to promote effective, all-hazard animal and agricultural emergency management. In addition to enhancing and clarifying communication, NASAAEP has now developed best practices for key animal issues that occur during emergencies (www.nasaaep.org). One of the nine Best Practice Working Groups[1] (BPWG) is Animal Search and Rescue. The role of that group is to identify those best practices associated with water- and land-based animal rescue.

HISTORY OF TECHNICAL RESCUE FOR ANIMALS

Technical animal rescue in the United States had a number of pioneers, but certainly, one of the leaders in this movement was Nicholas Gilman and the American Humane Association. Nick had responded to a number of disasters

[1] http://www.cfsph.iastate.edu/Emergency-Response/bpwg.php.

throughout the country starting with Hurricane Andrew and a number of disasters that required responders with technical skills including water- and land-based rescue. He was concerned that emergency services (fire and search and rescue (SAR)) were being tasked with handling difficult animal situations, and conversely, animal control officers were being put into situations that required technical rescue skills. What was needed was a bridge between human technical rescue and animal control.

In 1997, Gilman reached out to Rescue 3 International[2] to assist in developing the first course on Technical Animal Rescue (TAR).[3] The course was piloted in 1998 and, by the following year, being taught throughout the country by AHA.

At about the same time, there were a number of groups working on large animal rescue including Tomas and Rebecca Gimenez (SC), John and Debbie Fox (CA), and Roger Lauze (MA), so the late 1990s and early 2000s were an active time for developing animal rescue curricula.

In 2000, AHA recognized that the TAR course was not an ideal fit for their responders, and consequently, Dick Green and Gilman developed Water Rescue for Companion Animals and Rope Rescue for Companion Animals (WRCA and RRCA). Both of these courses were 28h in length and utilized multiple animal rescue scenarios and evolutions.

Hurricanes Floyd and Katrina illustrated the need for having well-trained water responders. By the time Katrina came onshore in 2005, there were a fairly large number of TAR and WRCA certificated folks, but very few of them had ever worked in a flood situation, and even fewer knew how to properly maneuver a boat or to capture animals while in the boat. Based on those experiences, Green and Gilman developed a slack-water rescue course and rolled it out as an IFAW course in 2010 (Fig. 1.4).

Following the Deepwater Horizon oil spill (April, 2010), Green (IFAW) secured a grant to improve the animal response capability for the Gulf States with an emphasis in Louisiana and Mississippi. As part of that grant, IFAW was able to provide equipment and course development for slack-water rescue to those two states. In 2011, Green joined the ASPCA Field Investigations and Response team and, in 2013, developed the current ASPCA course, Animal Slackwater Rescue.

The number of people impacted by disasters worldwide is increasing, and as shown in the following figure, we have seen a nonlinear increase in the numbers impacted in the last two decades. When there are people issues, there will be animal issues, and we must continue to include animals in our rescue, evacuation, and sheltering plans. Providing opportunities for families with animals to be colocated or cohabitated will ensure greater numbers evacuated and fewer folks attempting to go back into hazardous areas to recover their pets. Animals are a very important part of our culture and often treated as one of the family—it only makes sense to have well-trained animal rescuers in every community (Fig. 1.5).

FIG. 1.4 Nicholas Gilman and IFAW Water Rescue Team. *Courtesy of Nicholas Gilman and The International Fund for Animal Welfare. All rights reserved.*

[2] https://rescue3.com/.

[3] https://rescue3.com/tar/.

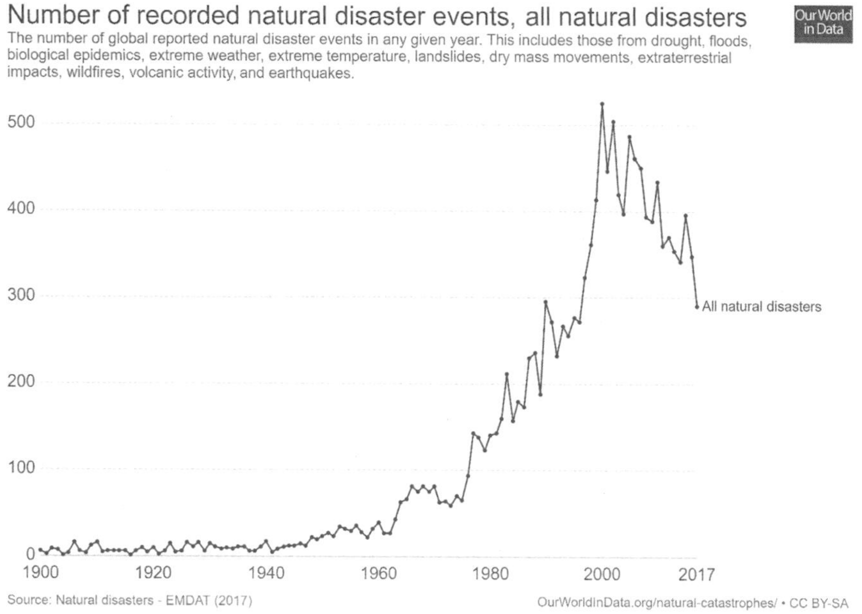
Number of recorded natural disaster events, all natural disasters
The number of global reported natural disaster events in any given year. This includes those from drought, floods, biological epidemics, extreme weather, extreme temperature, landslides, dry mass movements, extraterrestrial impacts, wildfires, volcanic activity, and earthquakes.
Our World in Data
500
400
300
200
100
0
1900
1920
1940
1960
1980
2000
2017
All natural disasters
Source: Natural disasters - EMDAT (2017)
OurWorldInData.org/natural-catastrophes/ • CC BY-SA

FIG. 1.5 Natural disasters 1900–2017.

RECENT LEGISLATION

Two key animal-related pieces of legislation were passed shortly after Hurricane Katrina. The Pets Evacuation Transportation Standards (PETS) Act[4] was signed into law in October of 2006 and amended the Stafford Disaster Relief and Emergency Assistance Act[5] to "ensure that state and local emergency preparedness operational plans address the needs of individuals with household pets and service animals following a major disaster or emergency."

The Post-Katrina Emergency Management Reform Act[6] (PKEMRA), also passed in 2006, recognized "colossal inadequacy" at federal, state, and local levels … to prepare for, respond to, and recover from large-scale incidents. The PKEMRA strengthened the Federal Emergency Management Agency's (FEMA) preparedness and response capabilities and identified new responsibilities for Department of Homeland Security (DHS) and FEMA in coordinating implementation of the PETS Act.

Hurricane Katrina and the subsequent passing of these two Acts along with the development of NARSC and NASAAEP have had a significant impact on the way that we respond to disasters today and the recognition of the importance of including animals in emergency planning. Following Hurricane Harvey in Texas (2017), it was found that the PETS Act and the lessons of Hurricane Katrina had contributed to a positive cultural shift to including pets (companion animals) in emergency response.[7]

RESOURCE TYPING

There have been a number of attempts at identifying and standardizing the roles and functions of animal first responders. Animal first responders are referred to as a *resource*. When a resource has the training, experience, and

[4] https://www.congress.gov/109/plaws/publ308/PLAW-109publ308.pdf.

[5] https://www.fema.gov/media-library/assets/documents/15271.

[6] https://www.doi.gov/sites/doi.gov/files/uploads/Post_Katrina_Emergency_Management_Reform_Act_pdf.pdf.

[7] Glassey S. Animals 2018;8(4), 47. doi:10.3390/ani8040047 Did Harvey Learn from Katrina? Initial Observations of the Response to Companion Animals during Hurricane Harvey.

equipment to effectively perform a designated task, that resource becomes a *capability*. The National Animal Rescue and Sheltering Coalition (NARSC) has identified five primary support functions that may be called upon in a disaster:

- Assessment
- Evacuation and transportation
- Search and rescue
- Sheltering
- Veterinary care

To perform one of the functions listed above will require trained and well-equipped personnel and, most often, teams of responders to complete the task. For example, sheltering requires a number of key personnel from a shelter manager to the behaviorist. Collectively, all of the identified roles can be combined into a sheltering team. The composition of that sheltering team may change as the size and type of shelter changes. A team tasked for managing a 50-animal shelter will look much different than a team tasked to manage a 3000-animal shelter. The process of defining and categorizing, by capability, the resources needed to complete a task is referred to as *typing*. Resource typing definitions establish a common language and define a resource's minimum capabilities. To ensure that a resource is capable of performing a designated task, there must be a process of identifying core competencies for each position on a typed team. *Credentialing* ensures that the identity and attributes of individuals or members of teams are validated against an established set of minimum criteria and qualifications for specific job titles.[8]

FEMA developed the National Incident Management System (NIMS) to establish a credentialing system and to "guide departments and agencies at all levels of government and nongovernmental organizations and the private sector to work together seamlessly and manage incidents involving all threats and hazards—regardless of cause, size, location, or complexity—in order to reduce the loss of life, property, and harm to the environment."[9] NIMS and resource typing will be discussed in greater detail in Chapter 2.

The process of having Animal Search and Rescue (ASAR) recognized as a legitimate response entity required buy-in from a large number of stakeholders as seen in Fig. 1.6. The goal was never to create an animal counterpart to human search and rescue but rather to offer a first step for building and enhancing existing animal response capabilities.

In 2015, NARSC developed ASAR Responder Certification levels (Fig. 1.7). The purpose of this undertaking was to identify those requisite trainings needed for every level of certification (awareness, operations, technician, and specialist). The goal of this proposal was twofold, to identify those trainings and/or experiences necessary to perform in the field (operations) and to set a goal for response groups to strive toward in having their members and teams recognized as qualified to perform field animal rescue activities.

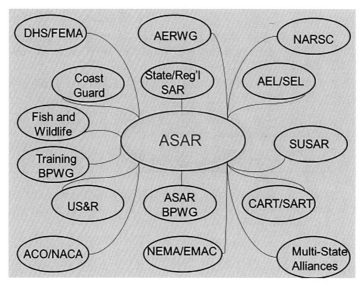

FIG. 1.6   Stakeholders in the development of ASAR.

---

[8] The NIMS Guidelines for Credentialing of Personnel: /national-incident-management-system/publications-index.

[9] FEMA NIMS: www.fema.gov/national-incident-management-system.

All certifications levels are proposed based on existing and/or potentially available courses and are recommended training levels.
Requirements are contingent for consideration of the AHJ and equivalents may be appropriate in some cases.

ASAR operations	ASAR technician	ASAR specialist
Reference FEMA 509-1. The ASAR Operations will be an Animal Care and Handling Specialist plus the following skills.		
ICS 100, 200; IS 700, 800[3]	ASAR Operations certification	ASAR Technician certification
IS 10, 11[3]	ICS 300	ICS 400
Animal Control, Capture and Behavior (NACA) (8 h)[1]	Swiftwater Rescue (24 h)	Experience as Team Leader/Command[4]
Awareness Level HAZMAT course	High/Low Angle Rope Rescue (24 h)	OSHA HAZWOPER (24 h)[11] or higher
Animal Handling Experience[2]	Awareness Level of Large Animal Rescue (8 h)	Operational Level Large Animal Rescue (24 h)
Introduction to SAR [3,10]	Introduction to Assessment (4 h)[5]	Wilderness First Responder (WFR)[8]
Animal Emergency Sheltering[6]	Fundamentals of SAR (FUNSAR) (47 h)[11]	Advanced SAR (ADSAR) (19 h)[11]
Human First Aid and CPR	Wildfire Operations for the Animal First-Responder[5] or higher	
Pet First Aid[9]	Compromised Structure Rescue[5]	
Safe Boat Handling (USCG approved)[3]	Wilderness First Aid (WFA)[7]	
Slackwater (16 h)	OSHA HAZWOPER (8 h)[11] or higher	
Wildland Fire Safety Awareness for the Animal First-Responder [5] or higher		
Animal Decontamination[5]		

[1]Or minimum of one year as ACO.
[2]Minimum of 1 year in shelter/clinic setting.
[3]Available on-line.
[4]Task Book.
[5]Course in development.
[6]NARSC-endorsed.
[7]Wilderness First Aid (NOLS WMLA – 16 h) or equivalent.
[8] Wilderness First Responder (WFR) or equivalent.
[9]ARC or equivalent training.

[10]SARTECH III or equivalent.
[11]Or equivalent.

Credential process may be done at four different levels:
  (a) CART recognized by County Emergency Management
  (b) SART recognized by State Emergency Management
  (c) State or Local Government Emergency Management Agency
  (d) National Animal Rescue & Sheltering Coalition (NARSC)

FIG. 1.7 Proposed ASAR responder certification levels.

In August of 2015, NARSC and NASAAEP hosted a boot camp in Louisiana to commemorate the 10 years since Hurricane Katrina. Seven courses from the Operations Level of the Certification list that follows were offered, and most of the courses were developed and taught by the members of the NASAAEP Best Practice Working Group (BPWG). The boot camp was by invite only, and the leading animal responders in the country participated along with members from state and federal government. It was an amazing example of the type of collaboration that now exists between the national NGOs, states, and the federal government.

FITNESS GUIDELINES

Animal Search and Rescue consists of periods of low activity punctuated by periods of intense, strenuous activity. Add the additional weight of personal and capture equipment along with the heat retaining properties associated with dry suits and you have the ideal conditions for creating physical and physiological stress. Good physical condition is a critical component in the body's ability to deal with stress and to perform the task successfully and without injury.

Medical and fitness standards have not been established for Animal Search and Rescue (ASAR) positions. However, the duties of such positions require good fitness. Until standards are established, the ASAR BPWG suggests following the guidelines for medical and fitness standards that have been established for wildland firefighters.

Wildland firefighters, much like ASAR responders, will engage in activities that call for above-average endurance and superior conditioning. There will also be times when the activity is "light" and calls for average endurance and conditioning. A number of work capacity tests (WCT) have been established to ensure that responders assigned to various activities are physically capable of performing their duties. The WCT is a family of tests to determine responder physical capabilities at three levels:

(1) *Arduous*: Duties involve fieldwork requiring physical performance calling for above-average endurance and superior conditioning. These duties may include an occasional demand for extraordinarily strenuous activities in emergencies under adverse environmental conditions and over extended periods of time. Requirements include running, walking, climbing, jumping, twisting, bending, and lifting more than 50 lb; the pace of work typically is set by the emergency situation.

(2) *Moderate*: Duties involve fieldwork requiring complete control of all physical faculties and may include considerable walking over irregular ground, standing for long periods of time, lifting 25–50 lb, climbing, bending, stooping, squatting, twisting, and reaching. Occasional demands may be required for moderately strenuous activities in emergencies over long periods of time. Individuals usually set their own work pace.

(3) *Light*: Duties mainly involve office-type work with occasional field activity characterized by light physical exertion requiring basic good health. Activities may include climbing stairs, standing, operating a vehicle, and long hours of work, as well as some bending and stooping or light lifting. Individuals almost always can govern the extent and pace of their physical activity.

Each performance level has its own unique work capacity test (pack test):

Performance level	Field test	Time required
Arduous	3 mi walk, 45 lb pack	45 min
Moderate	2 mi walk, 25 lb pack	30 min
Light	1 mi walk	16 min

The ASAR BPWG recommends that all frontline responders be able to pass the "arduous" test. These are responders who will be in full gear and executing rescue efforts. The committee recommends that staging personnel, boatmen, and responders not in dry suits or performing rescue pass the "moderate" test. Supervisors and personnel not actively engaged in rescue or sheltering efforts should be able to pass the "light" test. The ASPCA does not require its responders to meet these guidelines.

2

Incident Management

The one thing that nearly all emergencies and disasters have in common is initial chaos. Along with the normal confusion, frustration, and possible danger that accompany an emergency, there exists a response mechanism that can add to the hectic situation as numerous agencies and resources arrive at the scene. The ability of the individual, community, or federal response teams to effectively manage a disaster or emergency is referred to as incident or emergency management. Incident management is a process that utilizes planning, organization, coordination, and operational strategies in an attempt to control an emergency or disaster.

A potential emergency occurs whenever local rescue resources are dispatched. Generally, an emergency can be effectively dealt with in a timely fashion. A house fire or small brush fire would be considered an everyday emergency by most response agencies. An emergency becomes a disaster when local resources are overwhelmed and the event lasts for some time.

Examples of recent disasters involving a significant animal response include the LA flooding (2016), Hurricane Matthew (2016), Hurricanes Harvey, Irma, and Maria (2017) and Hurricanes Florence and Michael (2018). Generally, disasters are catastrophic in nature, wreak extensive damage, may cause large numbers of fatalities, and place an overwhelming demand on local and national resources.

It's important to remember that these definitions are relative to the individual who is directly involved in the event. For example, a single home destroyed by floodwater may be a normal or routine response for emergency management teams, yet to the family who lived in that home for their entire life and who lost everything that they owned and cherished, the event was truly a disaster.

Most disasters occur and are normally addressed initially at the local level. The community's response to the event will be determined by the level of preparedness, training, and effectiveness of their Emergency Operations Plan (EOP). An EOP is essential and provides information and actions to be taken so that residents, property, and animals will be protected in natural or technological emergencies.

An EOP should be developed in concert with animal control agencies, veterinary services, and the various shelters and businesses that might deal with animals on a day-to-day level and will likely be involved with animal issues if an emergency occurs. Many states and counties have organized Animal Response Teams that are activated during an emergency. These response teams are generally trained animal rescuers who are actively engaged in the community and are recognized by emergency management.

If the scope of the emergency is such that additional community resources are required, an Emergency Operations Center (EOC) will become operational. In some counties and states, the EOC has permanent staff for daily operations. In other areas, the EOC is activated at some point in the response phase of an emergency when additional coordination of resources is needed.

Local emergency management agencies are responsible for establishing priorities and policies for their community. During a disaster, the EOC is responsible for ensuring that these priorities and policies are adhered to and that adequate resources are available if a disaster occurs. The EOC is generally located at a fixed facility and houses representatives from the various department heads, government agencies, and volunteer disaster groups. Bringing all of the major "players" together in a single facility lends itself to a coordinated, well-planned response.

The role of the EOC is to ensure that the community's Emergency Operations Plan (may also be referred to as a Comprehensive Emergency Management Plan or CEMP) is carried out.

The EOC is responsible for the following:

- Ensuring that each involved agency is providing situational and resource status information
- Establishing priorities between incidents
- Acquiring and allocating resources required by incident management personnel
- Coordinating and identifying future resource requirements
- Coordinating and resolving policy issues
- Providing strategic coordination

EOCs can be organized in a number of ways, but as a minimum, their functions include the following:

- Coordination
- Communications
- Resource dispatching and tracking
- Information collection, analysis, and dissemination

It's oftentimes the EOC where ASAR teams first check in and get credentialed. It's important that ASAR teams be recognized by emergency management regardless of the agency requesting their assistance. It's highly recommended that whenever ASAR assistance is requested that the assisting team requires an official request from the Authority Having Jurisdiction AND the local emergency management agency.

NATIONAL RESPONSE FRAMEWORK AND EMERGENCY SUPPORT FUNCTIONS

In 1979, the Federal Emergency Management Agency (FEMA) was created and was given the authority to provide for all disasters regardless of cause. FEMA incorporated the elements of the Incident Command System (which will be discussed later in this chapter) into a federal emergency management plan referred to as the Federal Response Plan, which has since been changed to the National Response Framework (2008; rev 2013 and 2016).

The National Response Framework (NRF) is a guide to how the nation responds to all types of disasters. It is built on scalable, flexible, and adaptable concepts identified in the *National Incident Management System*[1] to align key roles and responsibilities across the nation. The NRF describes the principles, roles and responsibilities, and coordinating structures for delivering the *core capabilities* required to respond to an incident and further describes how response efforts integrate with those of the other mission areas.

The National Preparedness Goal (NPG)[2] describes 32 activities, called core capabilities, that address the greatest risks to the nation. The NPG organizes the core capabilities into five mission areas:

- *Prevention*: Prevent, avoid, or stop an imminent, threatened, or actual act of terrorism.
- *Protection*: Protect our citizens, residents, visitors, and assets against the greatest threats and hazards in a manner that allows our interests, aspirations, and way of life to thrive.
- *Mitigation*: Reduce the loss of life and property by lessening the impact of future disasters.
- *Response*: Respond quickly to save lives, protect property and the environment, and meet basic human needs in the aftermath of a catastrophic incident.
- *Recovery*: Recover through a focus on the timely restoration, strengthening, and revitalization of infrastructure, housing, and a sustainable economy, as well as the health, social, cultural, historic, and environmental fabric of communities affected by a catastrophic incident.

There are a number of core capabilities that correlate directly with ASAR as it relates to mass care services, mass search and rescue operations, public health, and situational assessment. When attempting to strengthen animal response capabilities through trainings and field guides, it's important to keep this national directive in perspective:

...partners are encouraged to develop a shared understanding of broad-level strategic implications as they make critical decisions in building future capacity and capability. The whole community should be engaged in examining and implementing the strategy and doctrine contained in this Framework, considering both current and future requirements in the process.[3]

---

[1] https://www.fema.gov/national-incident-management-system.

[2] https://www.fema.gov/national-preparedness-goal.

[3] FEMA 3rd ed. National Response Framework—Executive Summary.

As mentioned previously, the NRF provides context for how the *whole community* works together and how response efforts relate to other parts of national preparedness. Core to the NRF are the Emergency Support Functions (ESF), each representing the resources typically needed in times of a disaster:

ESF 1 Transportation
ESF 2 Communications
ESF 3 Public Works and Engineering
ESF 4 Firefighting
ESF 5 Emergency Management
ESF 6 Mass Care, Housing, and Human Services
ESF 7 Resources Support
ESF 8 Public Health and Medical Services
ESF 9 Urban Search and Rescue
ESF 10 Oil and Hazardous Materials Response
ESF 11 Agriculture and Natural Resources
ESF 12 Energy
ESF 13 Public Safety and Security
ESF 14 Long-term Community Recovery and Mitigation
ESF 15 External Affairs

There are three ESFs that deal directly with animals: ESF 6, Mass Care, Housing, and Human Services; ESF 8, Public Health and Medical Services; and ESF 11, Agriculture and Natural Resources.

The primary agencies for ESF 6 are FEMA and the American Red Cross. The primary purpose of ESF 6 is to coordinate and provide life-sustaining resources, essential services, and statutory programs when the needs of disaster survivors exceed local, state, tribal, territorial, and insular area government capabilities. As discussed in Chapter 1, we have seen several pieces of legislation that now require animals to be included in evacuation plans. As a result, we have seen a much greater willingness for communities to colocate/habituate humans and animals. The American Red Cross is a member of NARSC and FEMA participates in NARSC quarterly meetings, so the cooperation and collaboration between these groups have resulted in significant gains in the evacuation and sheltering of animals.

The lead agency for ESF 8 is the Department of Health and Human Services, and its purpose is to provide coordinated assistance to supplement state and local resources in response to public health and medical care needs following a significant natural or man-made disaster. Included in ESF 8 is the overall public health response and the triage, treatment and transportation of human victims of the disaster, and the evacuation of patients out of the disaster area as needed. One of the resources available in ESF 8 is the National Disaster Medical System (NDMS), which is a network between federal and nonfederal sectors with the intent to provide medical response, patient evacuation, and definitive medical care. The National Veterinary Response Team (NVRT) resides within the NDMS and will be discussed later in the guide.

The primary agencies for ESF 11 are the Departments of Agriculture and Interior. As it relates to animals, the primary purpose of ESF 11 is to control and eradicate an outbreak of a highly contagious animal/zoonotic disease and to provide for the safety and well-being of household pets during an emergency response or evacuation situation.

A well-coordinated response to animals in disasters requires considerable collaboration between federal agencies: FEMA, HHS, and US Department of Agriculture (USDA); their state counterparts (ESFs 6, 8, and 11); and national NGO partners. Through recent legislation, the leadership and initiative at the federal and state levels, and the networking made possible by NARSC and NASAAEP, tremendous advancements over the last ten years have occurred in enhancing animal response capabilities and ensuring communication, cooperation, and collaboration at all levels of a major response. Collectively, this has resulted in countless animal lives being saved.

NATIONAL INCIDENT MANAGEMENT SYSTEM (NIMS)

Note: This text assumes that users have completed relevant courses in incident management so the intent of this section is to provide an overview of NIMS and the relevant components of ICS.

At the core of emergency response is the National Incident Management System (NIMS). On February 28, 2003, President Bush issued Homeland Security Presidential Directive 5. HSPD-5 directed the secretary of Homeland Security to develop and administer a National Incident Management System. NIMS provides a consistent nationwide template to enable all government, private sector, and nongovernmental organizations to work together during domestic incidents.

NIMS is a comprehensive, national approach to incident management that is applicable at all jurisdictional levels and across functional disciplines. The intent of NIMS is to be applicable across a full spectrum of potential incidents and hazard scenarios, regardless of size or complexity, and to improve coordination and cooperation between public and private entities in a variety of domestic incident management activities. NIMS is composed of six components that work together as a system to provide a national framework for preparing for, preventing, responding to, and recovering from domestic incidents. These components include the following:

- Command and management
- Preparedness
- Resource management
- Communications and information management
- Supporting technologies
- Ongoing management and maintenance

COMMAND AND MANAGEMENT

NIMS standard incident management structures are based on three key organizational systems:

- The Incident Command System (ICS), which defines the operating characteristics, management components, and structure of incident management organizations throughout the life cycle of an incident
- Multiagency Coordination Systems, which define the operating characteristics, management components, and organizational structure of supporting entities
- Public Information Systems, which include the processes, procedures, and systems for communicating timely and accurate information to the public during emergency situations

PREPAREDNESS

Effective incident management begins with a host of preparedness activities. These activities are conducted on a "steady-state" basis, well in advance of any potential incident. Preparedness involves a combination of the following:

- Planning, training, and exercises
- Personnel qualification and certification standards
- Equipment acquisition and certification standards
- Publication management processes and activities
- Mutual aid agreements and Emergency Management Assistance Compacts

Preparedness also includes being aware of potential hazards and risks that can occur in the region. A hazard is anything that has the **potential** to cause harm. A risk is the **likelihood** that a hazard *will* cause harm.

RESOURCE MANAGEMENT

NIMS defines standardized mechanisms and establishes requirements for describing, inventorying, mobilizing, dispatching, tracking, and recovering resources over the life cycle of an incident. Following Hurricane Katrina, the Gulf States were simply overwhelmed by the outpouring of support for the animals impacted by the storm. Truck after truck arrived at the Lamar Dixon Expo Center in Gonzales, LA full of mixed bags of dog and cat food, blankets, cleaning supplies, and even booties. It was overwhelming. There was no system in place to be able to effectively manage these donations, so donated goods ended up wherever there was space.

Following most larger disasters, the community will likely be faced with the challenge of what to do with all of the donated product. A common approach is to offer unused supplies to nonprofits in surrounding areas or to host a community fair where individuals affected by the event can come and receive donated supplies. There comes a time when free food, veterinary care, and supplies needs to cease. As vendors in the affected area start to open their doors to the public again, they should not have to compete with an emergency shelter. A critical piece to recovery is getting stores

FIG. 2.1 Community fair. *Sonoma Fires, 2017. Courtesy of The American Society for the Prevention of Cruelty to Animals. All rights reserved.*

open and people back to work, and when that happens, food and supply distribution needs to come to an end for those that are in a position to pay for that product or services (Fig. 2.1).

COMMUNICATIONS AND INFORMATION MANAGEMENT

NIMS identifies the requirements for a standardized framework for communications, information management, and information-sharing support at all levels of incident management:

- Incident management organizations must ensure that effective, interoperable communications processes, procedures, and systems exist across all agencies and jurisdictions.
- Information management systems help ensure that information flows efficiently through a commonly accepted architecture. Effective information management enhances incident management and response by helping to ensure that decision-making is better informed.

SUPPORTING TECHNOLOGIES

Technology and technological systems provide supporting capabilities essential to implementing and refining NIMS. Examples include the following:

- Voice and data communication systems
- Information management systems, such as recordkeeping and resource tracking
- Data display systems

Supporting technologies also include specialized technologies that facilitate ongoing operations and incident management activities in situations that call for unique technology-based capabilities.

INCIDENT COMMAND SYSTEM

Analysis of past responses indicates that the most common cause of response failure is poor management. Confusion about who's in charge of what and when, together with unclear lines of authority, has been the greatest contributors to poor response.

[4] https://training.fema.gov/emiweb/is/icsresource/index.htm.

The command and management component of NIMS and specifically the Incident Command System[4] (ICS) address these concerns. The first component of ICS is establishing the command function that is directed by the incident commander and his command staff, which is composed of the public information officer, safety officer, and liaison officer.

The Incident Commander (IC) is responsible for overseeing the entire incident. In many situations, the IC is the first trained officer on scene. This may be the first responding fire engine, patrol vehicle, or animal control unit. If the event involves multiple jurisdictions, a Unified Command structure may be established whereby representatives from a number of agencies will share responsibility for overall incident management. The term IC has been used loosely (and incorrectly) over the years as the IC represents the Authority Having Jurisdiction or his/her delegate. Animal welfare agencies such as a humane society may be an active player in the response but not have legal authority unless it has been delegated.

The Safety Officer (SO) is responsible for the overall safety of the operation and may bypass the chain of command if necessary to stop and correct any unsafe acts immediately. This individual needs to be advised by a safety expert on animal handling and livestock issues to ensure that the same safety standards employed in human search and rescue are in place for animal rescue operations.

The Public Information Officer (PIO) is responsible for managing media affairs including delivering official announcements and preparing media releases. The PIO should be familiar with the local, state, and federal plans and how they interconnect. Public information is critical to incident management, so it's imperative that a system for managing information be established early on in an operation. In large-scale incidents involving multiple agencies, a Joint Information Center might be established to ensure that all of the information that is going out to the public by the responding agencies is coordinated and consistent.

Given that most ASAR teams are not for profit, they will have a need to be able to effectively reflect their work on their agency's website. In some situations, especially with cruelty cases, information may need to be vetted before posting, and that is often the role of the PIO. In addition, media attention can go a long way to assisting with fundraising, and the PIO can help attract and manage media opportunities. It is imperative however that if you are ever approached by the media, you direct them to the PIO and do not provide an interview unless it has been cleared by the PIO. In disaster situations especially, it's critical that the individual with the proper training and understanding of the situation be the spokesperson for that agency.

The Liaison Officer (LO/LNO) communicates with all of the participating organizations and agencies to ensure that all responders and resources are identified.

The Agency Representative (AR) position is an individual assigned to an incident from an assisting or cooperating agency who has been delegated authority to make decisions on matters affecting that agency's participation at the incident. Agency representatives report to the incident liaison officer.

There are four support functions serving command: operations, logistics, planning, and finance/administration.

Operations is that section responsible for performing the actual response and/or rescue operation (tactical operations). This includes coordinated communications, safety supervision, and information. All communications are conveyed via the incident command post. The command post (CP) is the location from which direction, control, coordination, and resource management occur. There is only one CP, and it houses the IC, planning function, communication center, and all agency representatives (hopefully including animal welfare). The operations section chief may oversee branch directors if needed. Oftentimes, animal rescue will be placed under the rescue branch as a division or group.

The **planning** section is responsible for designing the "game plan" and ensuring that this plan is complete, realistic, and achievable. To accomplish this, the planning section chief may assign various units to gather and analyze data during an incident by performing assessments. One of those units is the Resource Unit (RU) who is responsible for recording the status of resources committed to the incident. This unit also evaluates resources committed currently to the incident, the effects additional responding resources will have on the incident, and anticipated resource needs.

Trained and nontrained volunteers began arriving by the carload within days of landfall after Hurricane Katrina, and there were limited resource management systems in place to coordinate the well-intentioned convergent volunteers. When an agency or a volunteer responds without a written request by an Authority Having Jurisdiction (AHJ), it is referred to as *self-deploying*. Responders that are either unable or unwilling to respond under an Incident Command System often create a host of issues for emergency managers. As a result of the number of self-deploying groups to Katrina (and later Rita), most jurisdictions today require a formal request for assistance and, in most cases, a mutual aid agreement in place prior to allowing an outside agency to enter into the jurisdiction. Self-deployment and rogue groups will be discussed in Chapter 11.

The **logistics** section has two primary functions: service and support. The service branch includes communication, food services, and medical care for the responding personnel. Support branch functions include arranging for adequate facilities, procuring supplies and resources, and servicing equipment.

The **finance** section chief is responsible for the financial management and accountability during an incident including authorizing expenditures, contracting with vendors, and arranging interagency sharing of resources.

ICS helps all responders communicate and get what they need when they need it. ICS provides a safe, efficient, and cost-effective recovery strategy. ICS has several features that make it well suited to managing incidents. These features include the following:

- Ability to expand or shrink as needed
- A single or unified command system
- Modular organization
- Consolidated action plans
- Common terminology
- Manageable span of control, —five to seven subordinates per supervisor
- Designated incident facilities
- Comprehensive resource management

ICS has been tested for more than 30 years and used for the following:

- Planned events such as races, parades, and even adoption events
- Fires, hazardous material spills, and multicasualty incidents
- Multijurisdictional and multiagency disasters, such as earthquakes and winter storms
- Human and animal search and rescue missions
- Biological outbreaks and disease containment (such as foot-and-mouth disease)
- Acts of terrorism

The ability to communicate within ICS is absolutely critical. Using standard or common terminology is essential to ensure efficient, clear communications. ICS requires the use of common terminology, including standard titles for facilities and positions within the organization. Common terminology also includes the use of "clear text"—that is, communication without the use of agency-specific codes or jargon. In other words, use plain English. Many of the ASAR teams in the United States come from the animal welfare side and do not use 10-code so to ensure good communications between all of the groups—use clear text.

Resources, including all personnel, facilities, and major equipment and supply items used to support incident management activities, are assigned to common designations. Resources are "typed" with respect to capability to help avoid confusion and enhance interoperability.

Maintaining adequate span of control throughout the ICS organization is critical. Effective span of control may vary from three to seven, and a ratio of one supervisor to five reporting elements is recommended. If the number of reporting elements falls outside of this range, expansion or consolidation of the organization may be necessary. There may be exceptions, usually in lower-risk assignments or where resources work in close proximity to each other. For example, in slack water operations for an animal response, the typical number on a team is three with one boat operator, one animal handler, and a team leader.

Common terminology is also used to define incident facilities, help clarify the activities that take place at a specific facility, and identify what members of the organization can be found there.

Incident facilities include the following:

- The incident command post (ICP)
- One or more staging areas
- A base
- One or more camps (when needed)
- A helibase
- One or more helispots

Incident facilities will be established depending on the kind and complexity of the incident.

Only those facilities needed for any given incident may be activated. Some incidents may require facilities not included on the standard list. The facilities key to ASAR are the ICP where you will check in, the base where you will have your equipment checked, the camp where you will sleep and eat (and hopefully shower), and staging areas either where you will be with your equipment waiting for deployment or where you will be setting up a temporary shelter for holding animals brought from the field and awaiting transport.

Incident action plans (IAPs) provide a coherent means to communicate the overall incident objectives in the context of both operational and support activities. Within the IAP, you will find the following:

- Where the ASAR team fits into the overall response (207—Organizational Chart)
- To which group you have been assigned (204—Organization Assignment)
- How you will communicate (205—Radio Plan)
- Your operational objectives (202—Incident Objectives)
- Where to go if you are injured (206—Medical Plan)

Consequently, this is an important document for the ASAR team to have. In some operations, the IAP can be absolutely voluminous, so you may have to pick and choose what you want to take with you in the field. The 201 form is the Incident Briefing and will provide a quick overview of what's happening for that day. It's important to remember as stated earlier that the national NGOs will seldom be the Authority Having Jurisdiction and therefore will not be the lead agency or assume command unless that responsibility has been delegated by the AHJ. If you were to dive into the ICS structure, you would find that the animal response group is typically assisting at some branch or unit level in the operation section.

Communication needs for large incidents may exceed available radio frequencies. Some incidents may be conducted entirely without radio support. In such situations, other communication resources (e.g., cell phones or secure phone lines) may be the only communication methods used to coordinate communications and to transfer large amounts of data effectively. Alternate forms of communication will be discussed in a later chapter.

Following Hurricane Katrina, the only way that animal rescue teams could communicate was through texting and even that form of communication was limited. There were several teams that had handheld radios, but the frequencies were different, and so, there was no ability for the various teams in the field to communicate to one another (interoperability). For the recent mudslides in Santa Barbara County (Jan. 2018), ASAR teams were dependent on cellphones as the county-assigned radios were not effective in the hills surrounding Montecito.

During the response to Hurricane Matthew (Oct. 2016), there were a large number of animal rescue teams being dispatched from a number of agencies, and keeping track of where everyone was and what their operational objectives were became a challenge. ICS principles can help an agency address accountability through the following:

- An orderly chain of command—the line of authority within the ranks of the incident organization
- Check-in for all responders, regardless of agency affiliation
- Each individual involved in incident operations to be assigned only one supervisor (also called "unity of command")

SINGLE INCIDENT COMMAND

When an incident occurs within a single jurisdiction and there is no jurisdictional or functional agency overlap, the incident should be managed by a single incident commander who has overall incident management responsibility. NIMS calls this single incident command.

Interestingly, animal control seldom assumes an IC role even though they may be the AHJ. In large-scale cruelty cases, the IC is typically from law enforcement and in disasters; the lead agency will typically be fire or law enforcement. The determination of which agency will assume the role of IC will be based on who has jurisdiction and the most experience (and financial wherewithal) to be able to manage the event and the EOP.

UNIFIED AND AREA COMMAND

In some situations, NIMS recommends variations in incident management. The two most common variations involve the use of Unified Command and Area Command. Unified Command is an application of ICS used when

- There is more than one responding agency with responsibility for the incident,
- Incidents cross political jurisdictions.

For example, a Unified Command may be used for the following:

- A flood that inundates a local humane society. In this incident, the fire department, emergency management, and animal control may participate in a Unified Command,
- A flood that devastates multiple communities. In this incident, incident management personnel from key response agencies from each community may participate in a Unified Command.

In the disaster arena, animal welfare often is part of an Area Command where there are multiple incidents that cross jurisdictional boundaries. Most flooding, fires, and hurricanes impact multiple jurisdictions. For example, the 2016 floods in LA impacted a number of parishes that required coordination of animal resources at a state or regional (area) level.

An Area Command is an organization established to

- Oversee the management of multiple incidents that are each being managed by an ICS organization,
- Oversee the management of large incidents that cross jurisdictional boundaries.

Area Commands are particularly relevant to animal emergencies because these incidents are typically

- Not site-specific,
- Not immediately identifiable,
- Geographically dispersed and evolve over time.

These types of incidents call for a coordinated response, with large-scale coordination typically found at a higher jurisdictional level. There are times during large disasters where animal rescue loses sight of the bigger rescue picture. The response to Hurricane Katrina (2005) is a great example of where the animal groups were so busy with rescue and sheltering that they really didn't have an appreciation of where those operations fit into the larger ICS structure. As mentioned previously, a common place for ASAR in the organizational structure is under the rescue branch. ASAR can be its own group/division or be part of a task force (unit) with the human rescue teams.

On large or wide-scale emergencies or even multiple jurisdictional seizure cases that require higher-level resource management or information management, a Multiagency Coordination System may be required. Multiagency Coordination Systems are a combination of resources that are integrated into a common framework for coordinating and supporting incident management activities. These resources may include the following:

- Facilities
- Equipment
- Personnel
- Procedures
- Communications

The primary functions of Multiagency Coordination Systems are to

- Support incident management policies and priorities;
- Facilitate logistics support and resource tracking;
- Make resource allocation decisions based on incident management priorities;
- Coordinate incident-related information;
- Coordinate interagency and intergovernmental issues regarding incident management policies, priorities, and strategies.

As mentioned previously, ASAR teams will need to check in with the EOC (or ICP) upon arrival. Part of that check-in process will include verifying personnel qualifications and certifications. The EOC or the requesting agency must ensure that the participating agencies' and organizations' field personnel possess the minimum knowledge, skills, and experience necessary to perform activities safely and effectively.

It's important that the ASAR team is utilizing equipment that is not only effective and humane but also in good working condition to perform mission-essential tasks. In many cases, the ASAR team will have their equipment checked for worthiness when they register with the EOC.

All disasters occur and initially managed at the local level. The first level of response to an emergency is with local resources: animal welfare groups and County Animal Response Teams (CARTs). Once local resources are overwhelmed, the local AHJ can reach out to partner agencies that they have agreements with. Neighboring communities/agencies may be called in to assist through interlocal or mutual aid agreements. In addition, the AHJ may have agreements with other local and/or national NGOs. Once all local resources and their partners are tapped or expecting to be tapped, the local EOC will coordinate a request on behalf of the AHJ for additional resources to the state. If the emergency warrants, the state may even declare a state of emergency for a county that frees up state resources such as the National Guard and the State Animal Response Team (SART). And if the disaster impacts a large number of people (or animals), the state can make a request to the president for a federal declaration. If that occurs, federal resources can be activated.

Federal resources for animal-related emergencies might include experts from the USDA, the National Veterinary Response Team[5] (NVRT), and/or FEMA. It may also include supplies and equipment or transportation and security resources. As an example, following Hurricane Sandy, three NVRT teams were activated for supporting animal needs in New York City. NVRT is part of the National Disaster Medical System (NDMS), and they come as complete teams ready to provide emergency veterinary support. To support emergency sheltering, the ASPCA put in a request to the NY City Office of Emergency Management who in turn submitted it to the State Emergency Management (Division of Homeland Security and Emergency Services). Once approved by the state, it was sent to FEMA, and the request was filled by NVRT. As mentioned in Chapter 1, we have come a long way since Katrina in identifying the process for providing national and federal support for major animal incidents. NVRT's availability came into question following Hurricanes Irma and Maria (Sept. 2017) when they denied several requests for support, so time will tell whether NVRT will be a recognized federal veterinary resource.

States can also reach out to other states during emergencies through what is referred to as Emergency Management Assistance Compact[6] (EMAC). This state-to-state borrowing system is recognized by all 50 states and approved by Congress so as to address issues in licensure and compensation. To date, there are not a lot of recognized state animal resources available for an EMAC request, but the list is growing as more CARTs and SARTs are being developed, trained, and outfitted. A number of EMAC requests were submitted and filled during the 2017 hurricane season. The biggest challenge was securing medical teams, but hopefully, as state veterinary reserve corps continue to develop, there will be more medical resources available through state-to-state borrowing.

A recent example might be the best way to summarize this chapter. Hurricane Isaac was a deadly and destructive tropical cyclone that came ashore in Louisiana on August 28, 2012.

Tropical storm-force sustained winds, with gusts well over hurricane strength, knocked out power to hundreds of thousands, while heavy rainfall led to flooding. Many dams along the coastline were briefly overtopped, though they did not break completely and were later pumped to prevent failure. Overall, Isaac caused $2.39 billion (2012 USD) in damage and led to 41 fatalities.

Louisiana Governor Bobby Jindal declared a state of emergency on August 26. Later that day, reports of exposed levees in Louisiana began surfacing from local news outlets. Crews were reportedly dispatched to cover the exposed dirt with heavy plastic and fill gaps in the levees. Mandatory evacuations were ordered for St. Charles Parish and for parts of Plaquemines and Lafourche Parishes. Four thousand National Guard troops were activated in the state. On August 27, President Obama ordered federal aid to Louisiana to supplement state and local response efforts.

In the state of Louisiana, the Governor's Office of Homeland Security and Emergency Preparedness (GOHSEP) is responsible for coordinating the state's response to a major disaster such as Isaac. The Department of Agriculture and Forestry (LDAF) is responsible for coordinating animal issues at the state level. LDAF has agreements with NARSC and a number of NGOs to support those efforts.

The Louisiana State Animal Response Team (LSART) was developed in 2004, but its first real test came during Hurricane Katrina. LSART is not a government entity, and oversight comes from a foundation associated with the Louisiana Veterinary Medical Association (LVMA). LSART has agreements with a number of parishes and with LDAF. In addition, LSART has agreements with several NGOs including the ASPCA. The ASPCA also has agreements with LDAF (sheltering and rescue) and with several parishes. Following Isaac, LSART requested the ASPCA to assist in preparing their response and to help with assessments following landfall.

In Louisiana, each Parish has an animal control division and an office of Emergency Management (OHSEP). The chief elected official for the parish is the Parish President and he is ultimately responsible for all response activities within his/her parish. So, during the initial response to Hurricane Isaac, it was the Parish President through his OHSEP that was coordinating the response efforts. Animal issues were being handled by the parish animal control and coordinated through OHSEP.

[5] https://www.phe.gov/Preparedness/responders/ndms/ndms-teams/Pages/nvrt.aspx.

[6] https://www.fema.gov/pdf/emergency/nrf/EMACoverviewForNRF.pdf.

St. John the Baptist Parish (SJBP) and Plaquemines Parish had significant flooding issues following Hurricane Isaac impacting both livestock and companion animals. Animal control for SJBP reached out to LSART very early on August 30 to assist them in evacuating their shelter as floodwaters were approaching dangerous levels. LSART and the ASPCA were at their shelter within hours with a team of responders and sheltering supplies and equipment. Prior to arrival, LSART and ASPCA checked in with SJB OHSEP. At that point, LSART and the ASPCA were under the direction of the SJB animal control and OHSEP.

The next day, LSART and ASPCA were tasked to start water rescue operations throughout the flooded areas of the parish. There were well over a hundred requests to rescue companion animals. The ASPCA sat down with OHSEP to determine the best way to organize the addresses and to start a search and rescue operation. The 100+ addresses were then put into sections of a parish map and an organized search pattern developed to ensure that all of the addresses were reached quickly and that all of the impacted areas were assessed and searched if needed. The ASPCA organized six boat teams composed of LSART and ASPCA staff, and by midday of the second day, all of the addresses were reached and animals accounted for. Since the shelter was not in a position to house a large number of animals and since waters were receding quickly, it was decided to feed in place whenever possible and to only take those animals out that needed immediate medical attention.

This is a classic example of how animal rescue can be extremely efficient and effective. Local animal control is the Authority Having Jurisdiction, and as they are overwhelmed or anticipate being overwhelmed, they reach out to their partners for support. During this operation, LSART and OHSEP were in daily contact with GOHSEP (State) and LDAF to ensure that they had the most current situational awareness and could anticipate any additional resource needs. In this case, the ASPCA was under the control of LSART who reported to animal control who was being directed by OHSEP and Parish President. This was a local event being handled by local resources and their partners.

If this operation had escalated—much like we saw in Plaquemines Parish over the next several days—then the parish may have reached out to the state for additional resources. The state would then have activated their resources (LDAF) and contacted those NGOs that they had agreements with, and if those were not sufficient to meet the needs, they could have reached out to other states through an EMAC to NARSC or to the federal government for federal assets.

Federal resources typically come with a price tag. Each state negotiates a cost share agreement with FEMA when a federal declaration occurs and typically that is 25%. With an EMAC request, the borrowing state picks up 100% of the cost, so states much prefer to see their counties (parishes) handle issues at their level whenever possible. Regardless of where the resources come from including the federal government, the local entity still "owns" the disaster and is responsible for overseeing the response efforts. They may opt to delegate their authority to another agency but even that does not remove them completely from the response process.

As mentioned earlier, NIMS is the cornerstone for emergency response and incident management in the United States today. There are a number of NIMS/ICS courses available either in a classroom setting or online.[7] Most Animal Search and Rescue teams require the following prerequisite courses:

- Introduction to the Incident Command Systems: ICS 100
- ICS for Single Resources and Initial Action: ICS 200
- National Incident Management System (NIMS) An Introduction: IS 700
- National Response Framework, An Introduction: IS 800
- Animals in Disaster: Awareness and preparedness: IS-10.a
- Animals in Disasters: Community Planning: IS-11.a

MANAGING ANIMALS IN DISASTERS

As mentioned previously, all disasters occur locally and are managed locally. Within every community or recognized geographic region managed by a form of government, there will be an Authority Having Jurisdiction (AHJ) responsible for the care of animals following a disaster. In most communities, the AHJ is animal control. Some cities/counties may opt to separate animal control from sheltering and even have different oversight for each function—for example, law enforcement may oversee animal control with public health overseeing animal sheltering.

[7] https://training.fema.gov/is/.

In smaller communities, animal control may be contracted out to another county animal control agency or to a nongovernmental organization such as a humane society. Or animal control responsibilities will default to law enforcement or during a disaster, to emergency management. It's important to note that even if an agency has animal control responsibilities, that same agency may not have animal rescue responsibilities following a disaster. And in large-scale disasters, the animal AHJ may get pulled away from animal responsibilities to address human issues.

The AHJ responsible for animals in disasters is typically requested to have a representative at the EOC. For small departments, that may not be possible, but having that animal representative at the center of response coordination ensures that animal issues are being heard and addressed in a timely way. The AHJ is oftentimes responsible for oversight of the animal evacuation, rescue, and sheltering branches. To be able to adequately cover those branches and subsequent divisions/groups and units, the AHJ will be dependent on local volunteers such as the CART and other local resources (Community Emergency Response Team (CERT) and animal welfare groups) and outside nongovernmental organizations (through mutual aid agreement) to assist.

CHAPTER

3

National Animal Response Capabilities

Including animals in emergency response plans in the United States is important, given that an estimated 56% of American households now have at least one companion animal, and American society places value on the lives of pets, seeing them as part of a family and holding emotional and sentimental value.[1] Because of the high value placed, it is critical that emergency response planning considers how residents will make decisions about their pets in the event of a disaster and can accommodate the expectations of pet owners. The need for strong animal response capabilities may also grow over time based on three expected trends: projections that the US human population will increase by approximately 40 million by the year 2030, projections that an increasing numbers of US residents will live in areas that are prone to natural disasters, and the possibility that the frequency and intensity of natural disasters may increase over time due to climate change.[2]

There has been some progress in emergency preparedness for animals over the last decade. In 2006, the US government passed the Pets Evacuation and Transportation Standards (PETS) Act that requires all community planning entities to have an evacuation plan that includes families with animals. Furthermore, many counties and states have established animal response teams, dedicated to preparing and planning for and responding to animal emergencies in their states and communities. As witnessed by responders during Hurricane Gustav in 2008 and Hurricane Sandy in 2012, state and local authorities advanced considerably in terms of the inclusion of pets in their emergency plans.[3]

Despite improvements over the last decade, there have nevertheless been indications that there is still a critical need nationwide to enhance animal response resources and capabilities at the state and local level. For example, a 2014 study of Vermont towns, conducted by the Humane Society of the United States (HSUS), found that less than half of officials reported having an adequate location for evacuation of people and pets, and only 10% of respondents reported having animal supplies on hand in their municipality.[4] In some areas of the country, there have also been anecdotal reports that response teams have dwindled in size and become largely inactive due to a lack of training and response opportunities.

In order to address the unmet needs for animal emergency preparedness in the United States, the emergency management community must first conduct a need assessment to understand the gaps in existing infrastructure and capabilities. To date, no systematic assessment has been conducted to determine the United States' level of preparedness for managing animals in an emergency and to determine which jurisdictions or regions have critical deficiencies. To address this knowledge gap, the American Society for the Prevention of Cruelty to Animals (ASPCA) designed and conducted the National Capabilities for Animal Response in Emergencies (NCARE) study.

The NCARE study was a cross-sectional descriptive survey of officials who oversee emergency preparedness in US states and counties. The survey was administered through the Internet and by telephone. The full survey content for state and county/city versions can be found on the ASPCA website.[5] In brief, the survey covered the following items: the presence of a State or County Animal Response Team (SART/CART) or equivalent organization, organization and membership of the SART/CART, activities conducted by the SART/CART in the previous 12 months, jurisdiction's

[1] Walsh F. Human-animal bonds II: The role of pets in family systems and family therapy. Fam Process 2009;48(4):481–99.

[2] van Aalst MK. The impacts of climate change on the risk of natural disasters. Disasters 2006;30(1):5–18.

[3] Burns K. Hurricane Gustav prompts responders to evacuate pets, https://www.avma.org/News/JAVMANews/Pages/081015f.aspx; 2008 [accessed 2 January 2016].

[4] Gelb E. Pets and planning: a survey of municipal emergency planning and preparedness in Vermont Humane Society of the United States; 2014.

[5] https://www.aspcapro.org/sites/default/files/aspca-ncare-manuscript-jhsem.docx.

active typed animal teams, equipment owned by the SART/CART, the presence and size of supply caches for managing small animals (dogs and cats) and large animals (horses and livestock), plans to allow people and their pets to shelter in one location, and perceived training or planning needs for the jurisdiction. Questions for state officials also covered the details of other organizations involved in emergency response: veterinary reserve corps (or equivalent), livestock organizations, private animal shelters, and university extension programs. The survey was pretested with five counties, and wording was adjusted based on feedback provided. The study findings were published in the Journal of Homeland Security and Emergency Management (August 23, 2017) and are available on the ASPCA website.[6]

All US states and all US counties and cities with a population of ≥1 million were selected for survey contact. In addition, a random sample consisting of 25% of the counties (or county equivalent) with a population of <1 million was selected.

The study received eligible responses from 49 states (98%), 33 of 49 large counties (67%), 506 of the 766 small counties in the random sample (66%), and 59 small counties not in the random sample that responded spontaneously.

Overall, 31 states (65%) reported having a State Animal Response Team (SART), while 48% of large counties and 23% of small counties reported having a County Animal Response Team (CART) (Fig. 3.1). When results were stratified by FEMA region, all regions had ≥50% of states who reported having a SART or similar state organization, except Region IX, in which none of the three responding states reported having a SART. All states (100%) in FEMA Regions II and VI indicated having a SART. In the previous year, ≥60% of CARTs and SARTs had met as a group and conducted some sort of training or exercises, but 19% of small county CARTs were not active at all. In free-text comments about CART activities, CARTs also spontaneously reported responding to an emergency in the previous year and conducting public outreach events. Among counties without a CART, 215 spontaneously reported relying on other local entities for emergency response (sheriff, humane society, ranchers, etc.), and 69 counties spontaneously reported relying on state or regional teams. Several of these counties emphasized that they rely on owners to manage their livestock and that personnel from these entities are typically well adept at self-evacuation.

Participants reported a variety of models for their CART's or SART's structure and membership. More than 80% of respondents reported having government agencies such as Department of Agriculture or Department of Health as members of their SART/CART. Having private nonprofit animal organizations as members was more common for SARTs (90%) than for large counties (44%) or small counties (52%). About 68% of counties with a CART reported hav-

FIG. 3.1 Proportion of jurisdictions with County Animal Response Team (CART) or State Animal Response Team (SART), by FEMA region. Shading indicates percentage of small counties (<1 million in population) that indicated they have a CART. Hatching identifies the regions where all states (100%) reported having a SART. N/A=not applicable because emergency activities are not typically organized at county level in Region I. *Courtesy of The American Society for the Prevention of Cruelty to Animals. All rights reserved.*

[6] https://www.aspcapro.org/resource/ncare-checklist-rate-your-disaster-readiness.

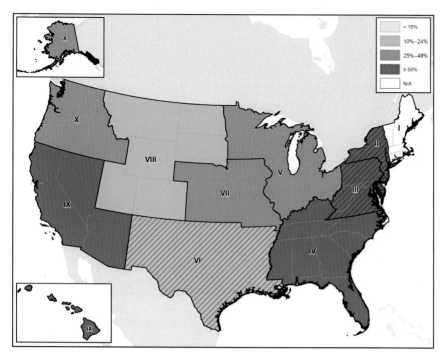

FIG. 3.2 Proportion of jurisdictions with a cache of small animal supplies, by FEMA region. Shading indicates the percentage of small counties (<1 million population) that indicated they have a cache of small animal supplies. Hatching identifies the regions where all states (100%) reported having a small animal supply cache. N/A=not applicable because emergency activities are not typically organized at county level in Region I. *Courtesy of The American Society for the Prevention of Cruelty to Animals. All rights reserved.*

ing individual members of the community as members. In free-text comments, many jurisdictions also described being part of multicounty or regional animal response teams.

A statistically significant, but weak, association was found between the historical number of declared disasters and the likelihood of having a CART. Even among those counties with a high frequency of past declared disasters (≥30 since 1953), however, only 30% reported having a CART.

Nationally, 74% of states, 77% of large counties, and 41% of small counties reported having a cache of supplies for managing small animals (dogs and cats) in an emergency (Fig. 3.2). Large animal supply caches (for horses or livestock) were less common with 52% of states, 38% of large counties, and 9% of smaller counties reporting having a cache of these supplies. When asked to estimate the number of animals that can be served by their caches, responses ranged from 1 to 1000 for large animals and up to 5000 for small animals. Among jurisdictions who could provide an estimate for their small animal caches, about a third (34%) of counties indicated that the cache could support at least 100 animals, and a similar proportion of states (37%) indicated that the cache could serve at least 250 animals. In free-text comments, 279 counties spontaneously reported using fairgrounds or public facilities for housing animals in emergencies.

Respondents were asked if their jurisdiction had plans to allow people and their pets to shelter in one location. In response, participants may have included collocated sheltering (CLS) and/or cohabitational sheltering (CHS). Only 50% of small counties reported having plans for collocated or cohabitational emergency shelters (Fig. 3.3), compared with 73% of states and 80% of large counties. In free-text comments, some counties reported limitations on companion animals in their shelters such as being open only to service animals in CHS.

Seventy-five percent of states have a veterinary reserve corps or similar organization, with the number of members ranging from 10 to 2000. Additionally, private nonprofit animal welfare organizations are engaged to provide support in emergency in 69% of states, while 89% of states indicated that livestock organizations are engaged.

More than 75% of respondents reported additional needs for emergency preparedness, such as training, expertise, and equipment. Training needs were identified by 84% of states, 65% of large counties, and 62% of smaller counties. A similar level of need was indicated for rescue equipment: 71% of states, 68% of large counties, and 62% of smaller counties. Subject matter expertise to assist with planning was listed as a need by approximately 45% of the jurisdictions regardless of type/size.

Most areas of the United States have, at the very least, taken some important initial steps toward establishing the capabilities necessary for managing animals in a disaster. In every FEMA region, for example, at least some of the counties and states have established a cache of supplies for sheltering small animals and have plans in place for collocated or cohabitational sheltering. Despite these indications of progress, many areas still exist for improvement. In

FIG. 3.3 Proportion of jurisdictions with plans for collocated or CHS, by FEMA region. Shading indicates the percentage of small counties (<1 million population) that reported having plans for collocated or cohabitational shelters. Hatching identifies the regions where all states (100%) reported having plans for collocated or cohabitational shelters. N/A = not applicable because emergency activities are not typically organized at county level in Region I. *Courtesy of The American Society for the Prevention of Cruelty to Animals. All rights reserved.*

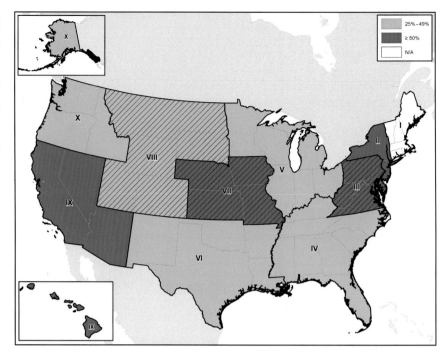

particular, the study findings point to some common deficiencies in county-level organization for animal emergency response. These deficiencies were particularly notable in FEMA Regions V, VI, VII, VIII, and X, where fewer than 25% of counties had a CART or similar organization. In contrast, FEMA Regions II and III were noteworthy for the high number of counties with a CART.

Among counties that would be expected to have the highest need for a CART based on a high historical frequency of disasters (more than 30 since 1953), only about a third have a CART. Even among the counties that have a CART, many appeared to be relatively inactive, as evidenced by having no meetings or activities in the previous year. The study demonstrated that jurisdictions with smaller human populations were less likely to have a CART or have taken other steps to prepare for animals in an emergency. In some of these counties, however, respondents reported in comments that the livestock owners take care of themselves and their neighbors in an emergency and do not expect or require any help from the county in an emergency.

FEMA has stressed the importance of enhancing emergency response capabilities by local government.[7] Organization at the county (or city) level is critical for emergency response to occur quickly enough to prevent animal emergencies. With emergent disasters—such as fire, flash floods, earthquakes, tornadoes, or chemical explosions—most animal emergencies occur within the first 24–48 h of a disaster onset (i.e., before state or national responders who can assist with animals typically have a chance to arrive). In those situations, rescuing humans will always be the first priority, with responders for animals typically attempting rescue only when it is safe and when it will not interfere with other rescue activities. It is not always possible, of course, to clearly separate the two different operations, especially when residents have pets and do not want to leave them behind. In that situation, having collocated or cohabitational shelters available and making their presence known to local residents could reduce the number of people at risk by increasing the likelihood that pet owners will evacuate early. As mentioned in the paragraph below on private animal organizations, one way that localities can prepare for animal rescue is to include, in the local emergency plan, private animal rescue organizations that can take on responsibilities for animal rescue as soon as such activities are safe, while other first responders are needed to be focused on rescue activities for humans.

Key steps toward county-level organization include the creation of a CART or a similar organization and cultivation of skilled, experienced, and motivated animal responders and include animals in the county's Emergency Operations Plan. In addition to establishing a CART, coordination with one or more adjacent counties (through, e.g., an interlocal agreement) is also a simple step for boosting response capabilities to localized emergencies. One benefit of taking these

---

[7] Pittman E. Remember: all disasters are local, says FEMA Deputy Administrator. Emergency Management; 2011. http://www.govtech.com/em/disaster/Remember-All-Disasters-Are-Local-Says-FEMA-Deputy-Administrator.html [accessed 14 November 2011].

steps is that counties with a CART, particularly if it is active and recognized by the local emergency management agency, are less likely to need outside assistance in an emergency, and response activities can happen quickly when they are more likely to save animal lives. Establishing a CART takes time, however, often requiring a year or more to establish key functions and approvals. Fortunately, the initial establishment of a CART does not require the governing body to first authorize its formation through legislation or changes of ordinance, minimizing one potential barrier to getting a CART started. For those counties wishing to develop a CART, there are guides available, such as Colorado State University's Community Animal Disaster Planning Toolkit.[8] Other resources to support CART development are available from the ASPCA and the US Department of Health and Human Services.[9] There are also a variety of models for organizing an entity that covers the functions of a CART. Some of the survey respondents, for example, reported successfully establishing multicounty CARTs or regional animal response teams.

About a third of states do not have a SART or similar organization. In general, the SART's function is not as a first responder. Instead, SARTs typically provide a critical role in coordination between counties and serve as a conduit for communication between counties and the state government. When a SART is lacking, intracounty coordination may be more challenging. SARTs also provide a critical role in coordinating the state response and in assisting counties with identifying and securing in-state resources including responders, equipment, and supplies.

The study found that only about half of counties with a CART had private nonprofit animal organizations as members of their CART. In contrast, the overwhelming majority of SARTs include private organizations as members. The responsibility for managing animals in an emergency may lie with a variety of different agencies, but in the immediate aftermath of a disaster, government agencies may be limited in their capacity to respond to situations involving animals. In some jurisdictions, for example, the local animal control agency alone does not have the staff capacity, equipment, or training to respond to an emergency, and government employees, regardless of role, may get dispatched for human rescue operations. Similarly, the local emergency management agency's priority is typically to save human lives. Therefore, many communities can benefit from identifying a community organization that can be charged with managing companion animals in an emergency. In order for them to respond quickly, however, agreements with these organizations (such as a *memorandum of understanding*) should be executed as part of planning activities, along with including official recognition of the animal organization in the local emergency plan and establishment of mechanisms for coordinating activities.

In the ASPCA study, about half of small counties lacked plans for collocated or cohabitational sheltering. In most cases, these shelters can be implemented effectively and safely, with little or no risk of adverse health consequences (such as allergic reactions) for shelter residents. They also cost substantially less than other models because the animal owners, rather than staff, provide care for the animals. It is important to remove barriers to evacuation, and experiences from Hurricane Sandy would suggest that residents were more likely to comply with evacuation orders when collocated or cohabitational emergency shelters were available, their presence was known to local residents, and pet-friendly transportation to the shelters was available. CARTs can serve an important function in working with local emergency management to address sheltering needs and ensuring that residents will have access to transportation to get to shelters with their animals.

The ASPCA study had a few limitations. The authors trusted that responding officials answered truthfully and to the best of their knowledge. Some respondents willingly shared their deficiencies in great detail, perhaps partly motivated by the hope that it would aid in garnering future support, but it is possible that others may have erred on the side of overreporting capabilities to avoid having their jurisdiction appear ill prepared. In order to minimize any tendency toward overreporting, the authors communicated that results would not be publically reported for individual counties. Another limitation was that the study was focused on state and county resources that could be easily quantified and reported in a survey format, but the authors were limited in their ability to capture expertise and capabilities of response teams and associated personnel. FEMA published the new Animal Resource Typing guide in June of 2018 so future assessments should be able to get a more accurate picture on stated capabilities though it will likely take some time to educate the counties and states on this process. The authors also assumed that responding counties represented US counties more broadly. They hypothesize that if the response was biased, that counties with limited animal response capabilities may be less likely to respond to the survey. In that case, the true levels of response capabilities for counties may be even lower than what they reported.

---

[8] Colorado State University Extension. Community Animal Disaster Planning Toolkit, http://extension.colostate.edu/disaster-web-sites/community-animal-disaster-planning-toolkit/; 2017 [accessed 1 February 2017].

[9] US Department of Health and Human Services. Animals in disaster, https://sis.nlm.nih.gov/enviro/animals.html; 2015 [accessed 1 February 2017].

4

Animal Resource Typing

Resource typing is defining and categorizing, by capability, the resources requested, deployed, and used in incidents. Resource typing definitions establish a common language and define a resource's (for equipment, teams, and units) minimum capabilities.[1] The effectiveness of a response is directly dependent upon the availability of resources. Resource typing expedites resource ordering and delivery thereby the effectiveness of response. All resources may be typed, but FEMA's animal resource typing includes nine animal response teams and 11 typed positions. Animal response equipment has not yet been typed. Resources are typed by level of capability; a Type 1 resource is generally considered to be more *capable* than Types 2, 3, or 4, respectively, because of size, capacity, experience, or qualifications.

Resources are identified by *kind* and *type*. The kind of resource describes what the resource is. For example, animal control vehicle, veterinary medical staff, portable X-ray machine, boat, and kennel cage are all *kinds* of resources. Kinds of resources can be as broad as necessary to suit the incident application.

The *type* of resource describes a capability for that kind of resource. Many tactical resources, such as animal control vehicles, will have a wide variety of capabilities and uses. If the Operations Section Chief ordered an animal van (kind of resource), the resource delivered may be inadequate. For this reason, it is strongly recommended that the various kinds of resources used be "typed" whenever possible. Animal control units in Florida in July need much better airflow and cooling than a similar situation in Western Washington.

Resources can be identified as follows:

- *Single*: An individual, a piece of equipment and its personnel complement, or a crew or team of individuals with an identified work supervisor that can be used on an incident.
- *Task Force*: A combination of single resources assembled for a particular tactical need with common communications and a leader.
- *Strike Team*: A specified combination of the same kind and type of resources with common communications and a leader.

Credentialing is a process whereby the identity and attributes of individuals or members of teams are validated against an established set of minimum criteria and qualifications for specific job titles. The *NIMS Guideline for the Credentialing of Personnel*.[2]

The first attempt at typing and credentialing animal resources came in 2007 as part of FEMA's National Integration Center's (NIC) development of the Animal Emergency Response (AER) Working Group. Over the next 2 years, this group developed a resource typing and credentialing framework to better prepare for and respond to all hazard causes of emergencies affecting all animals.[3]

In 2013, the Southern Agricultural and Animal Disaster Response Alliance (SAADRA) developed the Animal Emergency Response Typing Guidelines. In 2015, NARSC collaborated with SAADRA to revise those guidelines, and shortly thereafter, the guidelines were accepted by NASAAEP. The guidelines were vetted in regional exercises in

---

[1] https://www.fema.gov/resource-management-mutual-aid.

[2] It can be found at FEMA's website: https://www.fema.gov/media-library-data/20130726-1825-25045-1228/nims_guideline_for_the_credentialing_of_personnel_2011.pdf.

[3] https://www.fema.gov/media-library-data/20130726-1621-20490-3016/aer_credentials_for_nic_oct07_final_2_.pdf.

2014 and 2017 (Multi-Jurisdictional Animal Coordination Exercise) and served as the basis for the most recent guidelines published by the National Integration Center in 2018.

The NIC categorizes resources by team (NIMS 508) and by position (NIMS 509). In June of 2018, the NIC published the following list of animal response teams:

- Veterinary Medical Team (Types 1–4)
- Animal Shelter Team—Cohabitated (Types 1–4)
- Animal Shelter Team—Collocated (Types 1–4)
- Animal Shelter Team—Animal Only (Types 1–4)
- Animal Evacuation, Transport, and Re-entry Team (Types 1–2)
- Animal and Agriculture Damage Assessment Team (Types 1–2)
- Animal Search and Rescue Team (Types 1–3)
- Companion Animal Decontamination Team (Types 1–3)
- Large Animal Depopulation Team (Type 1)

The following NIMS 509: Position Qualifications were also published:

- Veterinarian
- Vet Assistant
- Animal Behaviorist
- Animal Response Shelter Manager
- Animal Care and Handling Specialist
- Animal Reunification, Tracking, and Identification Specialist
- Animal Emergency Response Team Leader
- Animal Control/Humane Officer
- Animal Search and Rescue Technician
- Animal Decontamination Specialist
- Animal Depopulation Specialist

The response to Hurricane Maria in September of 2017 is a great example of how animal resource typing can assist in identifying and requesting animal needs. Hurricane Maria was the 10th most intense Atlantic hurricane in history and the most intense storm the world saw in 2017. Maria was the third of a trio of storms to hit the states and territories—following hurricanes Harvey and Irma. It was the 13th named storm, eighth consecutive hurricane, fourth *major* hurricane, second Category 5 hurricane, and the deadliest storm of the 2017 season.[4] Maria is also the 3rd costliest hurricane on record with a total of roughly \$100 billion in damages; only hurricanes Katrina (2005) and Harvey from earlier in the season inflicted more damage.

Maria reached Category 5 strength on September 18 just before making landfall on Dominica. After weakening slightly, Maria achieved its peak intensity over the Eastern Caribbean with maximum sustained winds of 175 mph. On September 20, Maria struck Puerto Rico as a high-end Category 4 hurricane. The death toll from Maria will likely be higher than 1000 people in Puerto Rico that sustained the greatest loss of lives and significant destruction to infrastructure. Four months after landfall, there were still 250,000 people without power.

The US Virgin Islands also took a major hit from both Hurricanes Irma and Maria. On September 21 the ASPCA received a request from FEMA through the USDA to provide an assessment team for St. Croix. On September 26, the team arrived on the island, and on September 31, after meeting with all of the major stakeholders and interest groups, they provided their recommendations to the commissioner for the Virgin Islands Department of Agriculture (VIDOA). Their recommendations were based on the animal resource typing listed previously:

- One - Type 3 Companion Animal Sheltering Team
- One - Type 3 Large Animal Sheltering Team
- One - Type 3 Companion ASAR Team
- One - Type 3 Large Animal SAR Team
- Two - Type 1 Animal Behavior Specialists
- One - Type 4 Companion Animal Veterinary Team
- One - Type 4 Large Animal Veterinary Team

---

[4] https://en.wikipedia.org/wiki/Hurricane_Maria.

A link to all of the animal resource types can be found at the FEMA website[5] and in Appendix A. For this discussion, a Type 3 Companion Animal (Only) Sheltering Team and a Type 2 Animal Search and Rescue Technician will be examined.

A Type 3 Companion Animal (Only) Sheltering Team is requested when there are only animals versus a colocated or cohabitated shelter that will include owners. This would be a free-standing shelter with no owner interactions on a daily basis, so the sheltering team would have responsibility for the daily care of the animals in their charge. Following is the language for a Type 3 Companion Animal (Only) Sheltering Team.

Description: Responsible for the oversight, setup, operations, communication, and demobilization of a temporary companion animal shelter. The team is intended to provide a safe and protected environment for displaced companion animal populations ensuring their basic needs are met.

OVERALL FUNCTION

(1) Establishes and manages a temporary shelter for the safe and humane handling, care, husbandry, and housing of animals with one of the competency areas described below:

    (a) Companion animal: including pets, service, and assistance animals
    (b) Livestock: including food or fiber animals and domesticated equine species

(2) Ensures basic animal welfare needs are met.
(3) Ensures proper identification, tracking, reunification, and reporting of animals.
(4) Coordinates with Incident Command, all facets of the animal response, and intersecting components of the human response. Maintains safety and sanitation of the facility and equipment. Ensures appropriate security is provided.

COMPOSITION AND ORDERING SPECIFICATIONS

Discuss logistics for deploying this task force, such as security, lodging, transportation, and meals, prior to deployment:

1. This team works 12h per shift, is self-sustainable for 72h, and is deployable up to 14days.
2. Requestor may order a veterinarian, veterinary assistant, and veterinary medical team to support the team's activities, if necessary.
3. Requestor may order a Shelter Facilities Support Team Leader to support shelter operations, if necessary.
4. Animal-only shelters assume that full care of animals is provided by shelter staff (animals may be owned, but not with their owner or stray (owner not known)). Although owners may visit, they do not provide daily animal care. Average ordering ratio of Animal Care and Handling Specialists for animal-only shelter is 1 person to 15 animals, but this number will vary by species and sheltering conditions.
5. Consider sheltering requirements for animals that are medically or behaviorally not suited for congregate sheltering. Requestor specifies special management needs regarding species.

COMPONENTS

Component	Type 3
Total personnel	11
Management and oversight personnel	Same as Type 4 PLUS: 1—NIMS Type 2 Animal Response Shelter Manager
Operations and support personnel	Same as Type 4 PLUS: 1—NIMS Type 2 Animal Identification, Tracking, and Reunification Specialist

[5] https://rtlt.preptoolkit.fema.gov/Public/Combined?s=FemaId&p=7.

Capacity	Up to 100 animals
Location	Self-contained temporary shelter
Team equipment	Requestor will provide or obtain shelter kit to set up shelter based on animal population served, including the following:

1. Cages, crates, or other containment equipments
2. Leashes, halters, lead ropes, lariat ropes
3. Muzzles
4. Food
5. Potable water
6. Bowls
7. Litter boxes
8. Litter
9. Cleaning and disinfection supplies
10. Microchips
11. Vaccines

Personal protective equipment (PPE)	Same as Type 4
Communications	Same as Type 4

Following is the language for a Type 2 Animal Search and Rescue Technician.

Description: The Animal Search and Rescue (ASAR) Technician provides expertise in locating, capturing, containing, and evacuating of animals in postdisaster environment.

Overall Function: The ASAR Technician locates and rescues animals with one or more of the competency areas described below:

1. Companion animal: including pets, service, and assistance animals
2. Livestock: including food or fiber animals and domesticated equine species

COMPOSITION AND ORDERING SPECIFICATIONS

1. This resource can be ordered as an individual resource.
2. This resource can be ordered as part of a team (Animal Search and Rescue Team).
3. Discuss logistics for this position (such as security, lodging, transportation, and meals) prior to deployment.
4. This position works up to 12 h per shift, is self-sustainable for 72 h, and is deployable for up to 14 days.
5. Requestor specifies competency areas necessary based on the animal population the position will serve.
6. Requestor specifies specialty areas (e.g. fire, slack water, and swift water) necessary based on incident requirements.

COMPONENTS FOR TYPE 2 ASAR TECHNICIAN

Component	Type 2
Education	Not specified
Training	AHJ ASAR Technician Level training equivalent of NASAAEP Best Practices ASAR Training: Technician Level
Experience	Two years of experience in a disaster/emergency setting commensurate with the mission assignment
Physical/medical fitness	Same as Type 3
Currency	Same as Type 3 PLUS:
	1. Routinely responds to ASAR requests at local, regional, or national level
Professional and technical licenses and certifications	AHJ-certified ASAR Technician Level

NATIONAL QUALIFICATION SYSTEM

FEMA has developed the National Qualification System (NQS),[6] which provides a foundational guideline on the typing of personnel resources within the NIMS framework, plus supporting tools. The NQS describes the components of a qualification and certification system, defines a process for certifying the qualifications of incident personnel, describes how to stand up and implement a peer review process, and provides an introduction to the process of credentialing personnel.

In addition, FEMA has established the NIMS Job Titles/Position Qualifications, which define minimum qualification criteria for personnel serving in defined incident management and support positions.[7] The next step in the animal resource typing process will be to establish the minimum qualifications for all of the positions listed previously. At this time, there are Position Task Books whereby an animal responder can have his experiences logged and recognized for advancement. That is a project slated for 2019.

EMERGENCY MANAGEMENT ASSISTANCE COMPACT

Resource typing will allow for states to better define what animal resources they need. There is a process in place for states to borrow resources from other states and is called the Emergency Management Assistance Compact (EMAC).

EMAC is an all hazard-all discipline mutual aid compact that was signed into law in 1996 with all 50 states, the District of Columbia, Puerto Rico, Guam, and the US Virgin Islands enacting legislation to become EMAC members.

EMAC offers assistance during governor-declared states of emergency or disaster through a responsive, straight-forward system that allows states to send personnel, equipment, and commodities to assist with response and recovery efforts in other states. Through EMAC, states can also transfer services (such as shipping newborn blood from a disaster-impacted lab to a lab in another state) and conduct virtual missions (such as GIS mapping).

When a state requests assistance from another state through the EMAC process, the cost for those resources falls on the borrowing state or territory. A number of EMAC requests were made following Hurricane Maria. On St. Croix, the VIDOA requested small and large animal control teams and small and large animal vet teams. The vet teams were never filled, but they were able to secure one small and one large animal control teams through the EMAC process.

More information about what resources have been typed can be found at the Resource Typing Library Tool.[8] This is a FEMA website managed by the National Integration Center and allows for easy access to all of the FEMA-recognized typed resources.

MISSION READY PACKAGES

States will often pre-identify specific response and recovery capabilities that are organized, developed, trained, and exercised prior to an emergency or disaster. These are referred to as Mission Ready Packages (MRP). They are based on NIMS resource typing but take the concept one step further by considering the mission, limitations that might impact the mission, required support, the footprint of the space needed to stage and complete the mission, personnel assigned to the mission, and the estimated cost.[9]

The use of MRPs is a major component of the Emergency Management Assistance Compact (EMAC). Using the EMAC system, states can request or deploy equipment, personnel (including volunteers), and other resources through MRPs to assist in disaster recovery efforts. To develop MRPs, National Incident Management System (NIMS) resource typing is conducted by potential providers. Under EMAC, potential resources can be classified under one of the following categories:

- Animal health
- Emergency medical services
- Human services

[6] https://www.fema.gov/national-qualification-system.

[7] https://rtlt.preptoolkit.fema.gov/Public.

[8] https://rtlt.preptoolkit.fema.gov/Public.

[9] https://www.emacweb.org/index.php/mutualaidresources/emac-library/mission-ready-packages.

- Fire and hazardous materials
- Incident management
- Law enforcement
- Mass care
- Medical and public health
- Public works
- Search and rescue
- National guard
- Telecommunicator emergency response

Each state's emergency management agency oversees the development of potential MRPs that fit these categories or meet specific capabilities. In addition, each MRP includes additional information on the costs, limitations, space requirements, and more associated with the deployment and subsequent use of the MRP. This information allows requesting states to select the MRP that best fits their budgets, needs, and limitations.[10]

---

[10] http://www.astho.org/Preparedness/MRP-for-MRC/.

CHAPTER

5

Preparedness Activities

An integral part of helping people in disasters is helping their pets. This chapter will introduce a tool recently designed by the ASPCA to help a jurisdiction identify key preparedness activities, assess animal response capabilities, and identify ways to address gaps between preparedness and potential response.

HUMAN-ANIMAL BOND

The importance of animals in our society and in our lives was discussed in Chapter 1. The American Veterinary Medical Association (AVMA) notes[1] that due to a lack of more traditional support systems in modern society, for many people, companion animals are the sole source of emotional and social support, providing significant psychological and physical health benefits, especially to children, the elderly, the disabled, the mentally and physically ill, and the incarcerated. Given this bond, they believe that, "when disasters strike, saving animals means saving people."

Without question, these human-animal bond dynamics influence people's responses in disaster situations—causing them to behave in ways that put themselves, responders, and others at risk. In his paper "A study of pet rescue in two disasters"[2], Sebastian E. Heath found that up to an 80% reduction in premature reentry into evacuated areas could be achieved if pets were evacuated with their owners. He noted that of those who rescued their pets, 65% felt it was worth risking their life to do so and that having children and having more than one pet greatly increased the likelihood of rescue. Based on his findings, he strongly recommends the full integration of animal welfare groups and responders into emergency planning so they can be an integral part of all evacuations and responses and mitigate these risks. Providing care for animals during an emergency may facilitate the personal safety and care of a large segment of the human population.

FIRST-RESPONDERS SAVE LIVES

In disaster response work, most animal emergencies occur within the first 24–48 h of a disaster onset, and often, it is those local responders that are initially performing activities of search and rescue and sheltering. It typically takes state and national responders 24–48 h to arrive to assist with search and rescue activities. Enhancing animal response capabilities through an established, skilled, and actively engaged County Animal Response Team (CART) is recognized by emergency management and supports animal control efforts—saves lives. These responders provide an invaluable service to emergency management and to their community. Their unique resources and expertise in managing animal populations are critically important to governmental and nongovernmental agencies charged with responding to and managing human and animal populations in a disaster.

Animal response teams should recognize that rescuing humans will always be the first priority, with animal search and rescue occurring only when it is safe and will not interfere with human rescue activities. It is not always possible to clearly separate between human rescue and animal rescue operations, especially when communities do not include animals in their evacuation and sheltering plans. With appropriate planning and training, animal rescue organizations

[1] https://www.avma.org/KB/Resources/Reference/Documents/hab_and_disasters.pdf.

[2] International Journal of Mass Emergencies and Disasters, November 2000.

can safely assume animal rescue responsibilities, while other first responders are focused on rescue activities for humans. Oftentimes, they may work in tandem.

ANIMAL EMERGENCY PREPAREDNESS CHECKLIST

In 2016, the ASPCA developed a three-pronged approach to building animal response capabilities nationwide. The first step was to conduct a national survey, the National Capabilities for Animal Response in Emergencies that was discussed in a previous chapter. The purpose of the NCARE study was to determine strengths and weaknesses at the county and state levels. The survey concentrated on state and local animal response teams and their capabilities, as well as the availability and access to the necessary equipment and supplies needed to respond to animals in a disaster. The second step was to develop a preparedness checklist that would assist agencies in identifying best practices for developing response capabilities in their community. The final step was to provide assistance through grants, training, and subject matter expertise to requesting agencies located in strategic areas throughout the country that were seeking to enhance animal response capabilities.

The Animal Emergency Preparedness Checklist was developed after experiencing so many communities struggle to handle animal issues following a disaster. History has shown that jurisdictions that have a structure in place to address animals in distress typically experience far fewer challenges than communities that have not taken the steps toward addressing the complex animal issues that can come with an emergency. That concern was validated with the NCARE survey. Based on conversations with emergency managers about what their needs were and areas where they felt they fell short, the checklist was developed. To ensure that important areas of focus were not overlooked, the checklist was reviewed by nationally recognized subject matter experts and strategic partners. It was then tested in a number of small and large communities across the country.

The checklist can be used as an analysis tool, as a blueprint for improvements, or as a road map to building animal response plans. It will help identify where the community is well resourced and where additional support may be needed. The checklist is focused around six (6) key areas: essential infrastructure; organization and leadership; written plan elements; equipment and systems; rescue, sheltering, and reunification; and personnel, volunteers, and training. Agencies that are in the initial stages of planning for animals in emergencies can utilize the designated high-priority items as areas to begin. The entire checklist can be found in Appendix B.

An initial strategic area for a community to concentrate on is identifying and including planning partners that will be playing a role in the disaster, that is, fire, law enforcement, animal control, agriculture, animal welfare groups, and the organizations charged with human sheltering (e.g., Red Cross). By including these organizations in your initial discussions, the planning process and final outcome will be a collaborative effort and provide better awareness and buy-in from the related emergency response agencies. The written plan guides all of the elements of preparedness and response—mutual aid agreements, search and rescue, sheltering, equipment, supplies, and personnel—and the collaboration with partners in this process will ensure these areas are addressed.

Equally important to the written plan is establishing colocated/cohabitated shelter locations and equipment caches. The checklist suggests a cache of equipment and supplies to support sheltering for a minimum of 50 small animals and 50 large animals. The equipment necessary for small and large animals is vastly different. This is an area where planning partners with experience in small and large animal sheltering can provide valuable input.

Ensuring adequate animal response capabilities means recruiting, training, and equipping volunteers in the community to support the operation. Having an already identified agency or group to provide sheltering support is going to be central to rapidly standing up sheltering operations once evacuations begin and search and rescue operations are underway. The goal is to have the capabilities and resources to provide shelter and care for 50 animals for 72 h. This is where agreements with other counties, rescue groups, and national nongovernmental agencies are so valuable.

Animals are recognized as members of the family for many household in the United States today. Consequently, when disasters strike, emergency management must deal with human and animal issues. The Animal Emergency Preparedness Checklist was developed to provide communities with a method for identifying animal response strengths and weaknesses and to provide a road map for ultimately enhancing response capabilities. It will be most effective when the "whole community" is engaged in the process. The ultimate goal of the checklist is to build community resources and capabilities that will enable a community to effectively and safely manage animals in disasters, thus building a stronger, more disaster-resilient community. The checklist is available on the ASPCA website.[3]

---

[3] https://www.aspcapro.org/resource/ncare-checklist-rate-your-disaster-readiness.

DEVELOPING A COMMUNITY ANIMAL RESPONSE TEAM (CART)

First responders save lives. A well-resourced animal rescue team recognized and working collaboratively with emergency management under the direction of the Authority Having Jurisdiction will be an incredibly valuable asset when an animal emergency occurs. County Animal Response Teams (CARTs), like their human counterparts—Community Emergency Response Team (CERT), are typically composed of volunteers within the community that have an interest in helping out in times of need. The first step in developing a CART is to develop an animal planning team (APT). This group is a critical piece for communities starting from scratch or where animal planning has been slow in getting started. The APT must include representation from the Authority Having Jurisdiction (AHJ) and emergency management. Representatives from the various animal welfare groups, law enforcement, fire and emergency medical services, and the American Red Cross will ensure that all interests are involved. Having these groups on the APT at the beginning of the planning process will go a long ways toward ensuring buy-in when it comes time for developing the animal response plan and having the CART included in the Community Emergency Management Plan (CEMP).

The APT will be responsible for reviewing the animal-specific annexes within the CEMP to ensure that they reflect current animal response capabilities and that listed players are still active and willing to support the response efforts. The ASPCA's community preparedness checklist mentioned earlier may be helpful in making sure that the entire key animal planning is identified and addressed.

CARTs may take on all or parts of the animal response that includes assessment, search and rescue, transport, sheltering, reunification, and demobilization. How many of these functions they are involved with will be determined by their capabilities. Capabilities are acquired through specific training, proper equipment, adequate personnel, and community recognition and support.

Many CARTs across the country primarily do sheltering, whereas a number of CARTs specialize in small and/or large animal rescue. In fire-prone areas, CARTs are often used for rapid evacuations of large animals. Other than larger metropolitan areas, many CARTs will still be dependent on outside support to manage large-scale events. And that is why agreements with other counties and NGOs are so important. For large disasters like the US experienced in 2017 and 2018, there are few agencies or CARTs that can handle the demands that come with hundreds or even thousands of animals at risk.

TRAINING

Fig. 1.7 listed the suggested courses for individuals seeking certification for Animal Search and Rescue (ASAR). Many of those courses can be found online including the Incident Command System (ICS 100 and 200) and the independent study courses covering National Incident Management and the National Response Framework (IS 700 and 800, respectively). These four courses are required for many CARTs as the minimum requirements. For team leads, ICS 300 is recommended, and for those folks that might be asked to assist in a coordination center (EOC), ICS 400 might be helpful. Both of those courses are instructor-led, classroom courses. Other relevant online courses include the following:

- IS 10.A: Animals in Disasters: Awareness and Preparedness
- IS 11.A: Animals in Disasters: Community Planning
- IS 230.D: Emergency Manager: An orientation to the Position
- IS 5.a: An Introduction to Hazardous Materials
- IS 111: Livestock in Disasters
- IS 244: Developing and Managing Volunteers

In addition, there are two modules within the CERT training that are specific to animals.[4]

All of the courses listed above are great for developing an understanding and appreciation of how an incident is managed and who does what during a response. For a CART member, the most important training is perhaps basic animal handling and husbandry. Animal handling is more than the experiences gained from owning a pet. CART responders will be asked to handle unfamiliar and possibly frightened animals, and that requires handling skills not necessarily acquired with the family pet. Volunteering with the local shelter or rescue groups will help in developing those skills. In addition, there are NGOs that offer animal handling and behavior courses. Emergency sheltering will be addressed in a later chapter, but understanding the basic food and care needs of various species is an important part of the CART member training.

---

[4] https://www.fema.gov/media-library/assets/documents/27983.

FIG. 5.1 24′ Enclosed Animal Supply Trailer. *Courtesy of Lake County Animal Care and Control (CA).*

EQUIPMENT

The equipment needs for a CART will depend on the tasks that they are expected to complete. Search and rescue will require some advanced equipment for everything from high- and low-angle rescue to water-based rescue. If the team is going to do both small and large animal rescue, then they will need specialized equipment for moving large animals from difficult places. This type of equipment is expensive and requires ongoing maintenance, a secure place for storage, and a method for getting the equipment to the site.

Many CARTs use small utility trailers Fig. 5.1 for storing their cache of equipment and generally pull those trailers with a ¾ or 1T vehicle. The Companion Animal Mobile Equipment Trailer (CAMET), developed by North Carolina SART is capable of holding small animal rescue and sheltering equipment. The CAMET[5] unit comes completely stocked with everything a CART would need to set up a small shelter (see Appendix B). Those and similar units can run $14,000 fully stocked. If the team is looking for a new truck to pull the trailer, a new ¾T pick-up can easily run $50–60,000. Insurance, maintenance, licensing, and maintaining inventory can be overwhelming for many CARTs.

Large animal rescue requires more equipment and consequently takes up more space, but the size of a trailer seen here should still suffice. Some CARTs have included racks on the sides of the trailer to hold portable pens/panels for building temporary holding areas. A list of possible equipment needs can be found in Appendix D.

PERSONNEL

CART members generally share at least one common trait—a compassion for animals. There are a host of skills needed for a well-staffed CART team including search and rescue technicians, logisticians, planners, animal handlers, shelter managers, and daily care specialists. Folks with training in animal behavior, intake, and reunification will be valuable additions.

CART teams generally flourish initially in terms of recruiting new members—especially if there has been a significant disaster in the news or in the region. Maintaining that interest and keeping numbers stable are the real challenges for a CART. Ongoing trainings, deployment opportunities, and regular meetings are all ways to keep interest high and members involved. Most of the national NGOs provide opportunities for volunteering.

The ASPCA has a unique Response Partner Program[6] where CARTs and animal welfare groups can support the ASPCA when they respond to a disaster or animal cruelty case. This has become an extremely successful program with well over 500 members representing thousands of potential responders. Partners provide much-needed assistance by sending skilled staff and volunteers to assist with field rescue, crime scene documentation, and emergency sheltering. Partners also provide rescued animals a chance for a new life by accepting them into their adoption programs. ASPCA's response partners have placed thousands of animals from disaster and cruelty cases into new homes. Responders gain valuable experience and bring those experiences back to their community, thereby enhancing local response capabilities.

[5] http://sartusa.org/camets/.

[6] https://www.aspcapro.org/how-agencies-can-partner-fir.

COMMUNITY RECOGNITION AND SUPPORT

CARTs that are active members of the emergency management community will have much greater support when it comes time to deploy. That support can come in the form of assistance with logistics, supplies, equipment, situational awareness, and even liability protection for volunteers. If the CART decides to seek nonprofit status (501(c)(3)), they will likely find much greater community support for fund-raisers and outreach efforts when they are a recognized member of the response community.

As mentioned earlier, there are a number of places for CARTs within the overall response organizational structure. In most cases, they will fall under animal control, but there may be situations where it makes more sense for them to be a free-standing entity and operate as another Volunteer Organization Active in Disasters. In that case, they may be tasked with law enforcement, fire, or even emergency management. If the CART only provides sheltering support, they may be tasked to public health or directly to ESF 6, Mass Care. Regardless of where they are housed, it's critical that the CART never self-deploy and stay within their assigned tasks and jurisdictional area.

6

Monitoring and Activation

There are three categories of disasters as it relates to time for preparing: those that come with little or no warning such as an earthquake or explosion, those with limited warning such as a tornado and tsunami, and those where there may be days to prepare such as a winter storm or hurricane. There are a number of websites, applications, and television and radio weather channels that provide weather reports and short- and long-term forecasts. The National Oceanic and Atmospheric Administration[1] (NOAA) is a department within the Department of Commerce and has a threefold mission: (1) understand and predict changes in climate, weather, oceans, and coasts; (2) to share that knowledge and information with others; and (3) to conserve and manage coastal and marine ecosystems and resources.

The National Weather Service (NWS) is a branch within NOAA with "nearly 4,900 employees in 122 weather forecast offices, 13 river forecast centers, 9 national centers, and other support offices around the country to provide a national infrastructure to gather and process data worldwide."[2] In the last 20 years, there have been hundreds of private agencies that have entered the weather reporting and forecasting business—and it is indeed big business—a multibillion dollar business.[3]

The forecast process starts with data and observations that come from weather stations around the world, satellites, radar, reports from volunteers, and weather balloons that collect information about the atmosphere, such as humidity, wind speed, and temperature. All of this information is fed into supercomputers run by the US government and other countries. These supercomputers take initial conditions, use mathematical equations, and spit out a forecast. There are no perfect mathematical formulas (algorithms) simply because it's so difficult to gather so much information over such a large geographic area like the United States. So, at best, we can create algorithms that can make estimates based on the information provided. And since the computer models used are all slightly different and use different mathematical formulas, each forecast might be slightly different.

Weather applications are huge today and can be found for just about any mobile platform. NOAA/NWS provides weather data for free, making it relatively easy for someone to develop a weather app. The accuracy of the app depends on how the developer uses the information provided, and just because it has access to the information, it doesn't mean that the forecast will be accurate. For many of us, the forecast that we receive from television or a phone app is just fine for most situations. But for agencies and communities that require the most accurate forecast possible, they are probably going to use the NWS or private agencies that work closely with NWS. However, a number of private organizations have developed their own weather monitoring systems and proprietary computer models to generate their own forecasts. The more technology that these groups invest in generally equates to more frequent and more accurate forecasts.

Even with the best of technology, forecast accuracy may still be dependent on the skills of the meteorologist interpreting the data. The seasoned meteorologist will know the strengths and weaknesses of the various models available and will know which model will fit the weather situation the best (Fig. 6.1).

Many disaster-prone states will have dedicated meteorologists embedded within Emergency Management to provide situational awareness to state and county agencies and residents when inclement weather is forecasted.

---

[1] http://www.noaa.gov/.

[2] http://www.noaa.gov/our-mission-and-vision.

[3] http://www.forbes.com/sites/marshallshepherd/2016/06/07/when-it-comes-to-u-s-weather-forecasting-private-public-or-both/#3965c2494810.

FIG. 6.1   Forecasters in Norman, Okla. Providing the public with an early forecast of the May 31, 2013 tornadoes in Oklahoma. *Courtesy of NOAA.*

FEMA along with the 10 FEMA regions[4] provides daily updates on any potential weather-related hazards. These daily briefings are available to government and private agencies.

There are many websites and apps available, but for starters, the first responder might want to check out the following sites:

- During flood season, NOAA's website (http://water.weather.gov/ahps/forecasts.php)
- During fire season, the National Interagency Fire Center's site (https://www.nifc.gov/fireInfo/nfn.htm)
- In hurricane season, the National Hurricane Center, which is also part of NOAA (http://www.nhc.noaa.gov/)
- During tornado season in addition to the NWS, adding one of the tornado tracking systems such as the Weather Channel's Tornado Conditions (https://weather.com/tv/shows/amhq/news/tornado-torcon-index)

The first responder needs to keep in mind that weather conditions take time to analyze and distribute, so even weather channels with reporters on the ground may not have the most current or relevant information.

In addition to monitoring the weather conditions, the animal response community needs to have an awareness of what animals might be in the path of severe weather. Having well-established relationships with the animal welfare community will often be the best source for current information as it pertains to animals at risk. In those situations where an outside group (e.g., national NGO) has been requested to support response efforts, there are a number of local groups that will have the most relevant information about their animal population and what animals might be most at-risk.

- Local animal control and humane organizations. Typically, the group that is responsible for responding to animals in disasters is animal control. In many communities, they work closely with the humane groups, and in others, they don't—so some homework will be necessary before reaching out. Phone numbers and e-mail addresses are usually quite easy to find with a web search or you can try Petfinder (https://www.petfinder.com), which maintains one of the largest databases of shelters in the United States.
- Local emergency management. Once again, contact information is generally easy to find on the net. Keep in mind that all of the groups mentioned in this section will be extremely busy dealing with life-threatening situations so waiting a day or two to call about animals might yield a more responsive person on the other end.
  - In many communities and states, the CEMP is available on their website, so checking out who is the Authority Having Jurisdiction and reading the appropriate sections (ESF or Annexes) that deal with animals will go a long way with your success of contacting the appropriate agency.
- Local health department. In many communities across the United States, animal control resides with public health, and roles and responsibilities in an emergency may be found in ESF 8.
- Public Information Officer. Situational reports (sitreps) available to the public may have information about sheltering (human and animal) and will oftentimes have contact information for the PIO. Generally, these folks are very good at responding and relaying messages.
- Social media. Some folks might be surprised to see this option so late in a list. Facebook lights up whenever there is an emergency—especially if animals are at risk. There will be a great deal of valuable information from this source, but do not take what's posted as gospel. For as much good information you may find, there may be an equal amount of misinformation.

[4] https://www.fema.gov/fema-regional-contacts.

- Volunteer Organizations Active in Disasters[5] (VOAD). Members of the local VOAD (e.g., American Red Cross and Salvation Army) typically have firsthand knowledge of what's happening on the ground and may have relevant contact information and on-the-ground knowledge of animal situations. Following Hurricane Maria (2017), FEMA housed all of the responders on St. Croix on a cruise ship. The animal response team received extremely valuable intelligence from the various NGOs and for profit recovery teams.
- Local newspaper, television, and radio. Depending on the market size, you may be able to access current information about the event online as more media outlets post current stories online.
- State agencies. Departments of Agriculture, Emergency Management, and Health may have updates posted on their websites. State VOAD may have a website as well. During fire season, the state Forestry Department is generally quick in providing daily briefings and sitreps that may yield some good information on human and animal shelters that have been opened.
- Nongovernmental sites such as Cattleman's Association, universities, 4-H, Search and Rescue, Extension offices, and other animal-related groups will oftentimes have current animal situational awareness.

The Federal Emergency Management Agency (FEMA) has identified 10 regional offices, each headed by a Regional Administrator. The regional offices support the development of all-hazard operational plans and generally help states and communities become better prepared. These regional offices mobilize federal assets and evaluation teams to work with state and local agencies.

Each of FEMA's regional offices maintains a Regional Response Coordination Center (RRCC). The RRCCs are 24/7 coordination centers that expand to become an interagency facility staffed by Emergency Support Functions in anticipation of a serious incident in the FEMA region or immediately following an incident.

Each FEMA Region prepares a daily situation report, and it is a reasonably easy process for recognized response groups to get on their distribution list (Fig. 6.2).

While monitoring the situation, it's imperative that the team leader keep their team and agency aware of the situation. This gives folks a heads up—allowing some time to think about what needs to be done to meet their personal and professional commitments if they are asked to respond. It also allows some time for the agency to start thinking about priorities and potential resources and capabilities. The ASPCA uses the following levels of readiness when providing updates to the Field Investigations and Response team Fig. 6.3.

At some point in the monitoring and contacting phases, you may receive a request for assistance (Appendix F). If a signed agreement is NOT in place, then the supporting agency will want to take some steps to assure that they are not perceived as self-deploying:

1. Require a formal request from the AHJ and Emergency Management (EM). If the request originates with animal control and an agreement is not in place, then it is advisable to have the request originate with Emergency Management. This ensures communication between the AHJ and EM and gives Emergency Management a heads-up that you are responding and time to start the credentialing process.

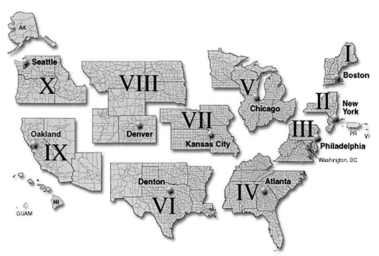

FIG 6.2 FEMA regions. *Courtesy of Federal Emergency Management Agency.*

[5] https://www.nvoad.org/.

1–4	Heads up	If you have close ties in the area, it may be worth a call to check in, but generally it's for information only
5–6	Readiness	A good time to make some calls to partner agencies to make sure that they are aware and taking steps to mitigate the potential impact. Checking for equipment and key staff availability
7–8	Alert	Time to have all calls made and availability determined from key staff/responders. Go-bag ready and plans in place in case you have to leave in a hurry. Good likelihood of deployment
9–10	Deployment	Very high likelihood of going out the door. On 24/7 standby

FIG 6.3　ASPCA levels of activation. *Courtesy of American Society for the Prevention of Cruelty to Animals. All rights reserved.*

2. Request a mission assignment (MA). In many communities, this will require that local Emergency Management request a tasking number from the state EM. If they use software (e.g., WebEOC), this is generally a fairly straightforward and quick process.
   (a) The MA should list the anticipated tasks, estimate for length of time needed, capabilities (typed teams) or types of resources, number of responders, point of contact, and start date and time.
   (b) If additional requests for services are made during the mission, it is advisable to request another MA.
3. Make sure the time is right for animal rescue. As mentioned previously, this can be a delicate and sensitive issue, but pushing animal rescue before adequate human rescue efforts have been completed not only is unprofessional but also may set back the animal response in the long run.

TEAM COMPOSITION

The decision on who the team leader will deploy will be based on a number of factors including availability, experience and training, credentialing, proximity, and physical capabilities. It's advisable to always put your best foot forward and bring the A-team when it's a big response, but remember that if the event goes on for any length of time, you will need to have strong leaders as teams rotate in and out. Typically, teams will operate on a 12 h duty period though it is not uncommon for the first-in team to pull a 24 h shift. Depending on the type of activities being performed, team members may need to rotate out after 7–14 days. High-stress environments such as working around hazardous materials will require a shorter rotation, whereas low-stress sheltering may be appropriate for a longer rotation. The biggest challenge for local rescue groups is what happens on Monday morning, and folks need to go back to work or need to go back to their regular jobs within the response agency. This is when mutual aid is so valuable.

Shortly after Irma made landfall in Florida, Lee County knew that they were going to need assistance and immediately activated their mutual aid agreement with a national NGO and they placed a mutual aid request for an ASAR team to assist in water rescue. That mutual aid request was filled by Miami-Dade Animal Services (MDAS), and within 24 h of the request, MDAS and State Fish and Wildlife were on the ground and in the water. MDAS also filled a mutual aid request from Florida Keys Humane Society to provide food and transport so mutual aid can be a valuable tool during a disaster. Hopefully, AHJs won't wait for a disaster to test the system and will engage in agreements with regional, state, and national partners. NIMS has provided guidelines for developing mutual aid agreements[6], and a template for a Mutual Aid Agreement is available on FEMA's website.[7]

The following guidelines may be helpful after the team has been identified and before they have been deployed:

(1) Provide a list of responders to EM so that they can provide credentials. That list should include name, role, and certifications.
(2) Be prepared to account for the activities of your responders. As mentioned in Chapter 2, the state may receive a federal declaration and have a 25% cost share. The number of volunteer hours and in-kind donations can be

[6] https://www.fema.gov/media-library-data/1510231079545-1fabc7af0e06d89d8c79c7b619e55a03/NIMS_Mutual_Aid_Guideline_20171105_508_compliant.pdf.
[7] https://emilms.fema.gov/IS706/assets/WyomingTemplate.pdf.

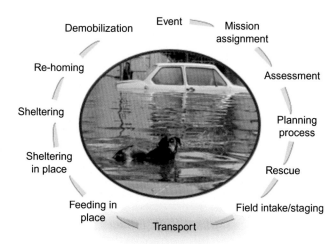

Circle of animal disaster work

FIG 6.4   Circle of disaster work.

applied toward that cost share. Check in with Emergency Management to make sure that the forms you are using for accountability are FEMA-approved.

(3) Select your responders carefully. If this is your first deployment in a community or state, put your best foot forward. Your responders will represent your agency, and it's critical that they do so in the most professional manner possible.

(4) Assign team and unit leaders. Once again, these are important roles and will likely be the only folks interfacing with the AHJ, so put some thought into who has the leadership skills needed and has the communication skills needed to enhance the response.

(a) Animal welfare can be a hot button in some communities. Your team leads and liaisons need to be aware of the sensitive balance between animal welfare and emergency response. An individual who can't see the herd through the individual cow may not be the best choice for a leadership role.

(b) Disaster response requires a cool head in hot times. A good leader can keep the big picture in their view even when nasty things are happening around them. These kinds of folks are hard to find and are best discovered in simulated stressful situations like trainings, scenarios, and exercises.

There is a host of activities that the responding agency may be asked to perform, and it's important that those tasks are clearly articulated on the mission assignment. The following figure highlights some of the response functions that may occur following a major event (Fig. 6.4).

CHAPTER

7

The Assessment Process

Assessing the level of damage and impact on animals is a critical part of the initial response phase. Animal rescuers—much like their human rescue counterparts—train countless hours to hone their rescue skills, so when the alarm goes off, the immediate response is to jump in a boat and go. But in the time that it takes to mount a local response, a rapid assessment could be conducted, and the information gained may save hundreds of animal lives. The information gathered from the assessment will drive the entire planning process and provide the intelligence needed for determining the scope of the response and the resources needed to ensure a safe and effective rescue mission.

An assessment answers three questions:

1. What is the extent of the damage, and how many animals have been impacted?
2. What are the critical short- and long-term impacts on animals?
3. What resources are needed to address animal issues?

To be able to answer these questions, the assessment team must have a basic understanding of the animal population and demographics *before* the incident, and that information will hopefully be available with local animal control or emergency management. Some communities are now incorporating animal interests (shelters, clinics, groomers, pet stores, etc.) onto community maps or utilizing geographic information systems (GIS) to better identify where animal interests reside. In this way, planners and responders can have a quick view as to where all of the animal interests are located and how to reach them. This makes the assessment process much easier and quicker. This information along with a general idea of the number of pets and backyard livestock known to be in an impacted area and half of the assessment's team duties are completed.

The assessment team (AT) has four primary objectives:

1. To assess the scope of damage within 24h of impact if possible
2. To assess the community and individual needs for addressing animal issues/concerns
3. To report findings within 24h of conducting the assessment
4. To recommend for the *type* and *capabilities* of resources needed to address animal issues

The report and recommendations provided by the AT must be clear, concise, timely, practical, and operational. This report will likely become the blueprint for how a community will develop their objectives, strategies, and tactics. A report that underestimates the needs may result in inadequate resources and unnecessary animal suffering. On the flip side, a report that overestimates the need may result in too many resources being requested and not being utilized wisely—let alone the cost involved in moving and managing those resources.

A key component in determining the needs for animals impacted by an incident is *vulnerability* or the ability of animals to cope. In the United States, rescue groups have felt the need to rescue every animal from an impacted area simply because it seems the right thing to do or the thing that might look impressive on their website. The reality is that many animals have the ability to cope moderately well following a disaster as long as they have access to food, water, and shelter. Consequently, feeding and sheltering-in-place programs may work very well for those animals that can be provided with resources and in situations where communities expect to recover quickly. Certainly, there will be times (long-term flooding, toxic spills, fires, etc.) when the animal needs to be extricated to an emergency shelter, but there are also many examples of when the animal will do just fine staying where it's at.

Vulnerability is dependent on a number of variables including *species*, *type of disaster*, *length of impact*, *availability of resources*, and *scope of impact*. It's important to note that *prioritization of species* by the AHJ may also have an impact on

FIG. 7.1 Japan Tsunami 2011. *Courtesy of The International Fund for Animal Welfare. All rights reserved.*

the ability of a species to cope. Wildlife is frequently not a high priority following natural disasters, but there have been some notable exceptions. Following the Deepwater Horizon oil spill[1] (2010), the number-one priority was the water-fowl given the impact on their habitat and the legal agreement between British Petroleum (BP) and the federal government. During the Black Saturday fires in Australia[2] (2009), the priority was for wildlife, and there have been a number of other disasters (2004 SE Asia tsunami—elephants) where wildlife or endangered species (2008 Chile Volcano—guemal deer) were the number-one priority, but generally, in most parts of the world, livestock is number one followed by companion animals.

Livestock are a critical part of the world economy and, regardless of where you respond, will likely be the number-one priority for the Department of Agriculture—the agency most likely to be in charge of the animal rescue efforts and the agency requesting *state* assessments. As you filter down to the local level, pets take on a much higher priority, and in many city/county jurisdictions, dogs and cats may be the most important species. Regardless of the group requesting the assessment or the priority species, the assessment team needs to be composed of subject matter experts (SME) that can address the needs for all species (Fig. 7.1).

An example might illustrate the difficulties that come when the AT does not have the species expertise needed to provide an accurate assessment of the needs. In 2000, a significant tornado hit a small Midwest community, and the local humane society (AHJ) requested an assessment team from a national NGO. They were able to tour the impacted area quickly and determine that there were no significant pet issues. They did however find a very large egg producer that was severely damaged. The AT did not have the expertise to determine the needs for how to deal with literally tens of thousands of dead and dying birds (Fig. 7.2).

A number of unrealistic options were proposed such as adoption or translocation, but when it came to depopulation, the AT was not prepared to deal with that option as it was in direct conflict with their philosophy and mission, and in addition to being such an emotionally difficult situation, the team did not have the specific training to realistically propose a response plan. There are similar situations on the opposite end where livestock assessment teams did not have the specific training/expertise for handling/planning for companion animal issues.

Is it the role of the AT to determine the best plan for dealing with a situation like this? No, but the AT needs to have enough experience with the impacted species to understand the scope of the impact and to help guide planners and the Resource Unit on potential resource needs.

As mentioned previously, it's important that the AT know all of the "players" involved in a response and their jurisdictional responsibilities. In the recent (2013) tornado in Moore, OK,[3] damage extended beyond the city lines and crossed into Oklahoma City. Consequently, animal control from the city of Moore and Oklahoma City were out rescuing animals. It was difficult to know where a found animal had originated and under whose jurisdiction it fell. Compounding the confusion was the response by the state and a large number of local and national rescue groups—none of which had legal responsibility for caring for animals from the impacted area. To be most effective

[1] https://www.epa.gov/enforcement/deepwater-horizon-bp-gulf-mexico-oil-spill.

[2] http://www.nma.gov.au/online_features/defining_moments/featured/black-saturday-bushfires.

[3] https://www.fema.gov/media-library/assets/documents/100807.

FIG. 7.2   2000 Johnstown Tornado. *Courtesy of American Humane. All rights reserved.*

in their job, the AT needed to know all of the players involved and their roles so that, hopefully, a coordinated response plan can be developed.

Shortly after the earthquake in Haiti, IFAW[4] and the World Society for the Protection of Animals (now known as World Animal Protection (WAP))[5] joined forces to lead the response efforts. They were able to meet with the Haiti Departments of Agriculture and Environment within days of the incident to map out a possible response strategy. Traveling anywhere within the impacted zone (air or land) was next to impossible, and the AT was dependent on the local veterinarian and NGO communities for information on the current (and past) situation. Based on the information gathered by the AT, it was determined to develop a mobile unit that would be able to respond to impacted areas and treat injured animals and provide free vaccinations. These teams reached over 100,000 animals over the 18 months that it was operational.

The AT was also charged with looking at the level of destruction to the core veterinary infrastructure throughout the country and what would be needed to get vets back into the field and providing valuable services including vaccinations. In this case, the AT was looking beyond the immediate damage and looking well into recovery and potential sustainability. This type of information would not be something that the human AT could provide to Department of Agriculture and why assessment teams with specific animal training—especially when it involves livestock and wildlife—are so important.

Interestingly, there have been numerous situations where providing vaccines was the only way that an NGO was able to access an impacted area. In countries where there is a higher rate of rabies or other zoonotic diseases, the AHJ oftentimes assumes that the only recourse following a major disaster where large numbers of animals have been impacted is depopulation. This practice is more prevalent in developing countries and where large numbers of human lives have been lost. Animal rescue groups have used a strategy of providing free vaccinations as a method for gaining access and sizing up the situation.

The Haiti response is a good example of the two *types* of assessments that the AT may be asked to do: situational (damage) and needs assessments. The situational assessment is a snapshot of the level of damage and scope or size of the impacted area. You will often see fire service personnel report on the number of acres burning and what percentage of the fire is contained. They are very accurate in their ability to make this type of assessment, and it is valuable *situational* awareness for the response communities.

The needs assessment is looking at what resources/services are needed by the impacted families and businesses to meet their immediate and long-term needs. These two types of assessments are done in concert with one another and typically at the same time depending on accessibility to the impacted areas. Following Hurricane Floyd, American Humane was requested to provide an assessment of the impact of the storm on veterinary clinics and animal shelters. Addresses were provided, and the two teams divided the area into sections. Over 2 days, the teams were able to visit all of the impacted sites and report back to the state (and owners in many cases) the level of impact and what it would take to get at least partially operational. As an interesting aside, as they were doing their assessment, the team came across

[4] http://www.ifaw.org/united-states.

[5] https://www.worldanimalprotection.org/.

FIG. 7.3   Indonesia, 2007. *Courtesy of The International Fund for Animal Welfare. All rights reserved.*

an abandoned shelter where the personnel had been ordered to evacuate without taking the animals. Water was literally flooding into the shelter as the team arrived, and they were able to move all of the animals to higher ground. This is not typically the role of the AT, but in this case, it was deemed an appropriate response and worthy of taking them away from their assigned task.

When an incident occurs, there may be a good deal of (*preliminary*) information available from a host of websites, media reports, and social networking. Calls to clinics and shelters, American Red Cross (ARC), Department of Transportation, and emergency managers will yield a good deal of information before even heading out to the field. Depending on the situation or the scope, the AT may be asked to provide an *initial* or *rapid* assessment. This can be done by vehicle (if accessible) or by air, and the information is relayed to command as quickly as possible. The rapid assessment is preferred when time is critical or when the size of the impacted area is larger. The *detailed* assessment is typically performed later in the response efforts and involves a door-to-door survey of the situation. The preliminary assessment can be finalized in less than a day assuming good access, communication systems are intact and the impacted area is relatively small. The rapid assessment may take 2 or more days depending on the area canvassed, and the detailed assessment may take a week or longer to complete.

In most disasters, time is of the essence, and the planning team will need the information from the AT as quickly as possible. That's not always easy, and rescue efforts may not be able to wait for that team's findings. Therefore, it's critical that the AT recognizes that the future of the operation may very well rest with the work that they do in the first couple of days following a disaster and the importance of their job cannot be underestimated. A number of years ago, IFAW was requested to assist in mudslides that had occurred in Indonesia. Numerous coastal villages were impacted over hundreds of miles, and the only road around the island was shut off for an undetermined amount of time. Getting a seat on a helicopter to do animal assessments wasn't going to happen for several days, so the AT rented a boat and went from village to village over 2 days to assess the animal impact. This allowed the team to see the damage up close, visit with the folks that stayed behind, and more importantly look at the condition of the pets and backyard livestock to determine the best plan of action (Fig. 7.3).

COMPONENTS OF AN ASSESSMENT

There are so many potential components that need to be addressed in an assessment that developing a form to capture them all is daunting in its own right. This section will attempt to identify some of the major components:

- *Area affected (location and size)*: Be realistic in the size of area that your teams assess. Multiple teams utilizing the same assessment techniques may be more effective and timely for large impacted areas.
- *Number affected (human and animal)*: Human numbers are important to know as it gives an indication of the potential pet population affected. Knowing the number of folks that are in shelters is also very helpful.

- *Mortality and morbidity rates*: High human mortality (initial Haiti estimate was over 250,000) is a good indication that there might be a large number of abandoned animals. Following Typhoon Nargis (Myanmar—2008), an estimated 200,000 water buffalo were lost, and the response community had a difficult time dealing with carcass removal. In situations such as the SARS epidemic (2002), there were large numbers of staff from shelters unable to report which had a significant impact on maintaining care for sheltered animals.

- *Types of injuries and illness*: Following Hurricane Sandy, the various shelters in the greater NY area were dealing with an 80% morbidity rate requiring extensive PPE requirements and specialized medical and sheltering procedures. Following the Tabasco flooding in Mexico (2007), the primary problems with equine and livestock were water-related sores and waterborne diseases as animals were stuck in the chest deep water for 10 days.

- *Characteristics and condition of the affected population (human and animal)*: This includes geopolitical, socioeconomic, cultural, and even religious considerations. There are sections of the United States that may place a higher priority on pets than they do on livestock or parts of the world where dogs and cats are not a welcome part of their community and even parts of the world where stray dogs and cats are captured and sold for food. All of this information is important to include in an assessment and will directly impact response.

- *Emergency medical, health, nutrition, water, and sanitation*: Availability of veterinarians and medical facilities will influence the triaging process. Moving animals out of an affected area is never an ideal situation as it greatly reduces the chance for rehoming, but when there is not adequate infrastructure in place to deal with injured animals, it may be advisable to transport out of the area.

- *Level of continuing or emerging threats*: Severe weather seldom moves through a region quickly, and during peak tornado season, it's not uncommon for severe weather threat to stay high for several days. This not only impacts the ability of the AT to complete their assignment but also may impact their recommendations. For example, during the Great Mississippi Flood (2008), priority was given to those areas where floodwaters were projected to stay high for the longest period of time. Changes in disaster and weather conditions will greatly influence where staging areas will be set up and where emergency shelters may be selected.

- *Damage to infrastructure and critical facilities (animal-related)*: Hurricane Harvey (2017) is a great example where local infrastructure was severely impacted—many of the animal shelters were flooded and without power and running water. This influenced not only the ASAR team response (limiting removal) but also transport (evacuating shelters). Knowing the status of local vet clinics will help the response team know when animals can be seen locally or where the closest operating emergency clinics are located. This information is critical for determining outside medical resources as well.

- *Damage to homes and commercial buildings*: Knowing the number of homes impacted will help determine the potential number of animals affected as many of the prediction formulas use number of "households." Unoccupied commercial buildings oftentimes are the place of choice for setting up emergency shelters, and knowing the impact on commercial buildings will give an indication of recovery time.

- *Damage to agriculture and food supply system*: It's always preferable to tap into local resources whenever possible. Bringing in a large amount of supplies from outside the area when it might be available locally slows down economic recovery. Every attempt should be made to use local vets, local feed stores, and local volunteers. The AT needs to report on how those local resources are faring and their ability to support operations.

- *Damage to economic resources and social organization*: Grocery stores, banks, ATMs, and restaurants are all critical parts of a community's ability to be self-sustaining and to be able to support response efforts. Following Hurricane Sandy, there were extreme gas shortages and lines that went for miles. All of the primary response vehicles for the ASPCA used diesel and had 75 gal reserve tanks in the bed that allowed them to work through most of the critical time without having to worry about fuel. As the response efforts continued, one of the most important pieces of intelligence was which gas stations were open and had the shortest line.

- *Vulnerability of the population to continuing or expanding impacts*: There is often a threshold for a community to preserve its ability to be self-sufficient after an incident. The second storm, additional rain, and even wind may affect that threshold, and the AT needs to have an awareness of impending weather/events and the community's ability to sustain another hit.

- *Level of response and capabilities*: What is a community's ability to handle the problems, and how dependent will they be on outside help? Some states are incredibly well prepared and have taken steps to reduce their risk to disasters. Other communities have taken very few steps to address animal issues following a disaster and will be completely dependent on outside resources.

- *Level of response from other donor counties, states, and NGOs*: This is an important piece of the assessment process, and it not only gives an indication of capabilities but also gives an indication on the level of regional and national involvement. A community that has an agreement with a national NGO and a state that has an agreement with NARSC would suggest a region that has spent time in preparing for and their ability to respond to a disaster.

An assessment is only a snapshot in time and information and conditions can change rapidly. The AT needs to look backward to capture the event as it evolved and lean forward to anticipate future needs. The *significance* of information can also change over time. What was important on Day 1 may not be significant on Day 7. And with animals specifically, what you don't see may be as important as what you do see. Following the SE Asia Tsunami in 2004, AHA responded to Sri Lanka to assist WSPA. The AHA lead was assigned to take the first team to the east side of the island and conduct an assessment. There had been reports of packs of dogs roaming the villages and the military shooting them to keep them out of the few food establishments that were up and running. The AT along with a small vaccination team agreed to conduct an assessment to determine the needs for the incoming teams.

Approximately 100 dogs and a handful of cats were visible to the team, and yet, the feedback from the community consistently reported larger numbers of dogs roaming the streets at night. Had the AT finished their work at that point, using the information from their first walk-through, they would have misrepresented the situation and that in turn would have affected the ability of the incoming teams to properly address the issues. On the second evening in town, the AT got up at two in the morning to tour the village, and there were literally hundreds of dogs roaming about. The AT needs to keep in mind that just counting what they see is not always an indication of what is out there and some of the challenges that might need addressing. *The assessment drives the objectives that drive operations.*

The information gained from the assessment will serve as a *baseline* for how the operation will progress. Therefore, the assessment report needs to be iterative rather than overly detailed and provide information in a timely way so that command, planning, operations, and logistics can get the response process started (Fig. 7.4).

There are a number of tasks that the AT will take on before starting their assessment including the determination of the known animal interests in the impacted area:

- Large confined animal feed operations
- Backyard livestock
- Medical/academic research facilities
- Veterinary clinics/hospitals
- Animal feed and supply stores
- Grooming facilities
- Boarding facilities

There are a number of sources for helping determine the potential number of animals in a community. Very few jurisdictions actually have a head count of the number of animals in their area, so most depend on prediction formulas,

FIG. 7.4   China earthquake, 2003. *Courtesy of Pixabay/Angelo Giordano.*

for example, AVMA 2012 survey[6] and the aggregate national statistics of the American Pet Products Association (APPA).[7]

Using prediction formulas is often concerning and may yield a high error of estimate. Pet ownership and populations vary with the type of community such as the following:

- Residents of large cities have fewer pets than those in smaller communities.
- There are higher rates of pets in single-family homes over apartments.
- Mobile home residents have the highest level of pet ownership.
- More affluent families are more likely to own pets.

For the purpose of this text, we will only consider estimating the number of pets, but there are a number of sources available for nonpets (exotics, backyard livestock, wildlife, etc.), and the AT will need to determine the populations of concern before starting their assessment.

One quick method for estimating pet population is to multiply human population by 59% (number of pets = 0.59 × total population).[8]

The type of assessment conducted will depend on a host of factors including amount of time, available resources, access, and prestorm populations. When the first response team arrives on the scene, the very first thing they do before "jumping into action" is conduct a *size-up* of the situation. This may be accomplished with a visual scan and a walk-around of the impacted area. For a single clinic or shelter, that might include determining the level of involvement, structural integrity, health and well-being of the housed animals, access, and resources needed to address the issues. In essence, a *size-up* is a quick snapshot and most often refers to the immediate assessment of a situation smaller in scope.

A *rapid* assessment is conducted when time is of the essence and a larger area needs to be assessed. Major components such as primary infrastructure damage, accessibility (ingress/egress), and major challenges are the norm in a rapid assessment. There is typically not the time for visiting individual shelters and clinics but rather a drive-by to get a general sense of the damage, resource needs, and recovery challenges.

Rapid assessments can be done by ground (often referred to as a windshield assessment) or by air. As mentioned earlier, animal issues do not typically earn a spot (at least early on) in the helicopter, but for situations like floods and fires, 30 min in a helicopter viewing the impacted areas can provide the same level of info as that of a 6 h drive, and oftentimes, the aerial perspective adds much to the big picture response plan.

Following Hurricane Katrina, ASAR teams were incredibly frustrated as water level and water conditions were changing rapidly and situational awareness was not available to the animal rescue groups.

After a week on the ground, AHA brought in one of their responders who had a helicopter and was able to gain permission to fly over the impacted areas (Fig. 7.5). That "intel" greatly improved the ASAR teams' ability to get into the right places at the right time! The aerial assessment team was able to identify all of the areas that still had significant floodwaters and identify the best access points.

FIG. 7.5 AHA helicopter, Hurricane Katrina, 2005. *Courtesy of American Humane. All rights reserved.*

[6] https://www.avma.org/KB/Resources/Statistics/Pages/US-pet-ownership-calculator.aspx.

[7] http://americanpetproducts.org/Uploads/MemServices/GPE2017_NPOS_Seminar.pdf.

[8] FEMA IND Committee.

The *thorough* assessment is conducted when time is not critical or when response teams are not waiting on key situational awareness. This is typically a door-to-door, block-by-block approach and will yield the greatest amount of information but generally takes longer than is available in the response phase of most disasters. However, in situations like the Deepwater Horizon oil spill, it took many days for assessment teams to be able to get into all of the coastal areas to assess the damage, but the information they were able to gather was critical for long-term recovery.

TABASCO FLOODS, 2007

The state of Tabasco in southern Mexico was hit with the worst flooding in its history in 2007 with nearly 80% of the state underwater. The town of Villahermosa was especially hard hit as floodwaters impacted livestock and pets alike. The IFAW assessment team arrived in Villahermosa on November 5, 2007. Over the next 3 days, the team performed an assessment of the general area through air, water, and land reconnaissance. The impacted areas included Villahermosa (pop 700,000), which is the capital of Tabasco (pop 2.2 million) and is situated on the low-lying swampy plains leading down to the Gulf of Mexico, and the remainder of the state of Tabasco that is largely agricultural (bananas, rice, beans, and corn).

The flooding was caused by a cold front that brought 5 days of torrential rain beginning on Sunday, October 28. Tabasco often suffers from flooding. After the last serious flooding in 1999, the state and federal governments began working on a complex flood-control project, but it was never completed. At the peak of the flooding, levels reached 19 ft. According to state officials, approximately 100,000 people were in shelters around the state, and over 800,000 people were homeless, but surprisingly, the death toll was very low.

Fortunately, IFAW had great connections with emergency management and agriculture through its relationships with the university (Universidad Nacional Autónoma de México), various response groups, and the veterinary community. It was those relationships that landed them a seat on a military helicopter within 24 h of arriving on scene. The impacted area was huge and rich agricultural lands and livestock areas were significantly impacted. Entire villages were stranded and dependent on helicopters for relief.

The AT was able to fly over the majority of the area and actually land at a village to bring in human and animal food. With this rapid assessment and intel gathered from local, state, and federal agencies along with discussions with the village leadership, the AT was able to complete their assessment and sit down with the state to discuss the response plan within 48 h of arriving.

From just the photo below (Fig. 7.6), the AT was able to determine the approximate number of households impacted and through discussions with local authorities was able to estimate the number and types of animals left behind or in shelters. That information was extremely valuable in determining the amount of food needed, prioritizing response areas, gaining access into affected areas, and calculating the number of rescue and feeding teams needed.

FIG. 7.6 Tabasco, Mexico, 2007. *Courtesy of The International Fund for Animal Welfare. All rights reserved.*

THE BIG DITCH PIG RESCUE, 2008

The Midwest section of the United States experienced a very warm winter and significant rain in the spring of 2008. By June 7, the Iowa, Cedar, and Mississippi Rivers had all crested resulting in large-scale evacuations and sheltering of humans and animals. By June 20, there were approximately 2500 companion animals in shelters throughout the region.

Flooding continued throughout Iowa for weeks, and floodwaters continued to move down the Mississippi River System where sections of Illinois also suffered significant damage due to flooding.

SE Iowa is a large pig farming and agricultural area. The town of Oakville is located along the Iowa River just before it empties into the Mississippi River. The river in this area was projected to crest on June 16, and large-scale evacuation orders had been issued. On June 14, the levee just outside of Oakville breached, and the entire town and surrounding area was under water in hours. For the next 2 days, human and animal rescue efforts took place. Within days, pigs were being spotted along the levee as those that were fortunate enough to escape their barns were swimming to the only piece of ground visible. As they would attempt to come out of the water, they would tear up the plastic barrier and sandbags protecting the remaining levee. The townspeople were worried that they would damage the levee to a point where additional sections of the levee would fail, so the Sherriff ordered his deputies to shoot any pigs observed on the levee.

On June 22, the Iowa State Department of Agriculture requested the assistance of IFAW who was working in concert with AHA and Farm Sanctuary, and an assessment team was dispatched on June 23 to survey the levee. Six teams were provided GPS coordinates that corresponded approximately with similarly sized assessment areas, and the teams traveled by boat to their assigned area and erected a pop-up tent with limited provisions. They would then begin their assessment on the levee until they reached the next team's pop-up tent. In this fashion, the six teams were able to cover over 25 mi of levee in less than 4 h and located (with GPS coordinates) nearly 70 pigs that were then brought to safety over the ensuing days (Fig. 7.7).

The form found in Appendix E may be useful as a template for your assessment team. It was designed by the NASAAEP Best Practice Working Group and may need to be modified to fit your agency or locale.

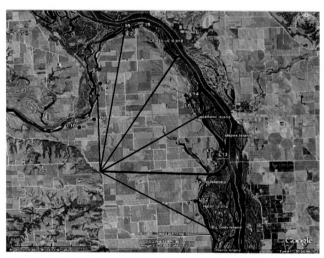

FIG. 7.7   SE Iowa flooding, 2008. *Courtesy of The International Fund for Animal Welfare. All rights reserved.*

CHAPTER

8

Evacuation and Transportation

This chapter will discuss the evacuation and transportation of household pets—primarily dogs and cats. The same principle applies, however, whether it be traditional pets or backyard livestock: the responsibility to safely evacuate animals lies first and foremost with the owner. It is very important to have an evacuation plan for all species including horses, livestock, exotics, research animals, and exhibition animals. Owners of nontraditional pets are encouraged to create a contingency plan for their animals; jurisdictions with any of these animals are, in turn, encouraged to engage these stakeholders in developing contingency plans.

According to the American Veterinary Medical Association, there are approximately 140 million household pets, nearly 5 million horses, 6 million livestock, and 13 million chickens in the United States.[1] A large number of those animals may be considered a pet and part of the family. The bond between humans and animals may be so strong that a family is reluctant to be separated from their pet even in times of emergency.

Recent disasters have demonstrated that people may not evacuate if they cannot take their animals with them. Issues also arise if the people do not have assurances that they and their pets will be colocated or cohabitated when seeking shelter. Those who are forced to evacuate without their pets may attempt to reenter the evacuated area before it is safe to do so to rescue their animals. In either case, this places a greater burden on the first responders that are tasked with the safety of people within the impacted area. Therefore, the need to be able to evacuate animals is paramount, and the ideal situation would be one where people and animals are evacuated together whenever possible.

While most people will likely be able to evacuate with their pets, some of the population will not have the resources to do so. Emergency management planners must consider this population to ensure that they and their pets and service animals will be included in emergency evacuation plans. Service animals must stay with their owners. The Department of Justice recently amended its regulation implementing title II rule of the Americans with Disabilities Act (ADA).[2] Included in those revisions was a change in the definition of a service animal:

A dog that has been individually trained to do work or perform tasks for the benefit of an individual with a disability.

The rule states that other animals, whether wild or domestic, do not qualify as service animals. Dogs that are not trained to perform tasks that mitigate the effects of a disability, including dogs that are used purely for emotional support, are **not** service animals. The final rule also clarifies that individuals with mental illness, a history of seizures, or post-traumatic stress disorder, among other nonphysical disabilities, who use service animals that are trained to perform a specific task are protected by the ADA. The rule permits the use of trained miniature horses as alternatives to dogs, subject to certain limitations. To allow flexibility in situations where using a horse would not be appropriate, the final rule does not include miniature horses in the definition of "service animal." If responders are unsure if an animal is a service animal, only two questions can be asked:

- Is the animal required because of a disability?
- What tasks or a service has the animal been trained to perform?[3]

☆ Taken in part from NASAAEP Evacuation and Transportation BPWG: DeSousa, Poirrier, and Green

[1] AVMA 2012 U.S. Pet Ownership & Demographics Sourcebook.

[2] http://www.ada.gov/regs2010/factsheets/title2_factsheet.html.

[3] 28 C.F.R. § 35.136(f).

EVACUATION PLANNING

Evacuation planning focuses on maximizing the number of people evacuated from a dangerous area, preferably prior to an incident. Orderly evacuation of animals concurrent with human evacuation will eliminate owners' resistance to evacuate or to abandon an animal before the incident and remove a postincident motive to return to an unsafe incident site. Thus, planning for animal evacuation promotes public safety during incident response and recovery.

All pet owners should be encouraged to prepare an emergency evacuation kit for their household pets and service animals that includes documentation of vaccinations especially rabies, a list and a supply of the pet's medications, a supply of the pet's food, and a pet collar that has the pet ID tag with owner's name and contact information and current rabies tag. Pet owners should have a transport carrier for their pet that is small enough to be put in a vehicle and a sheltering carrier big enough for the pet to stand up in and turn around. A complete list can be found at the ASPCA website.[4]

Pet owners should also be encouraged to permanently identify their pet with a microchip and register the pet with a national database. Microchipping is the most widely accepted form of permanent identification for pets. Identifying pets with microchips requires a scanning device. Most microchip companies provide tags that include the microchip number and the company's toll-free number, providing an additional method of identification if a scanner is not available.

One striking example of how well microchips work is the example of the horses removed from New Orleans and the surrounding area after Hurricane Katrina. Over 400 horses were removed from the area after the levees were breached. In Louisiana, horses are required to be permanently identified either with a microchip, brand, or lip tattoo when tested annually for equine infectious anemia. Because all of the horses were permanently identified (the majority with microchips), over 95% were reunited with their owners. This is in marked contrast with the household pets rescued after the levee breeches. The majority of household pets were not microchipped and registered to their owners. According to the Louisiana Society for the Prevention of Cruelty to Animals (LASPCA), the highest estimates were that 15% of household pets were reunited with their owners.

SHELTERING OPTIONS

Evacuation and sheltering plans go hand in hand as you must have a place to take people that are evacuated. The number and type of sheltering available will have an impact on the evacuation process. There are a number of emergency animal sheltering options including commercial housing, emergency shelter for animals only, colocated shelters, and cohabitated shelters. Animal boarding facilities and pet friendly hotels should be identified in sheltering communities, and a list of these facilities should be maintained by the authority having jurisdiction. This list should be updated annually. Emergency animal sheltering will be discussed fully in Chapter 12.

Cohabitated Facilities

Cohabitated shelters (CHS) is a relatively new construct to emergency sheltering and are designed to keep the entire family together—humans and pets—in the same area. Floor space is assigned to a family, and they are responsible for keeping their animal secured in that designated space and when taking the animal to outside areas. This type of sheltering has now been tested in a number of places following an event, and the advantages to colocated and emergency freestanding shelters will be discussed later.

Colocated Facilities

As with human shelters, colocated shelters (CLS) should be identified in sheltering communities as an option for people who evacuate with their pets. Use of such shelters should be encouraged if other options are not possible and publicized as such. One of the main limiting factors in encouraging people with household pets to evacuate is the limited number of sites accepting both people and pets. Having an adequate number of colocated shelters with equipment and staffing and informing the public about these shelters will encourage people to evacuate.

Colocated pet shelters may not be immediately adjacent to the human population shelter. In such cases, transportation to and from the shelters may need to be arranged.

---

[4] https://www.aspca.org/pet-care/general-pet-care/disaster-preparedness.

ASSISTED EVACUATION

For people that do not have a method of transportation out of the evacuation area, a plan will need to be developed that will address the special transportation needs of the community and their pets. *Every effort should be made **not to** separate pets from their owners*. Keeping people and pets together decreases the person's and the pet's stress and is less labor-intensive.

Experiences from recent hurricane evacuations demonstrated that limited resources to transport people may make it necessary to have separate transport of pets that cannot sit on the lap or fit in a carrier under the seat of an owner. During Hurricane Gustav (2008), large pets were separated from their owners at the local parish collection sites. Each pet was identified with a unique animal ID that linked the pets to the owner. Large pets were then placed on pet transport trucks to be transported separately from their owners. Forty-two percent of the total pets evacuated were large pets transported in this manner.[5]

There may be times when pets will need to be separated from an owner with medical needs. Pets may not be allowed in medical needs shelters, and it may be impracticable to set up colocated pet shelters near every human medical special needs shelter. A prison in Louisiana[6] has been designated to have inmates shelter and care for the pets of the medical special needs human population. Inmates participating in prison labor have been trained to assist in the setup, cleaning, and breakdown of pet shelters across the state, and they do it very well.

DEFINE ROLES AND RESPONSIBILITIES

As discussed previously, the Authority Having Jurisdiction (AHJ) is the entity ultimately responsible for animal issues. The responsible entity at the local level should be familiar with the jurisdiction's laws and ordinances pertaining to animals, animal ownership, animal bites, and dangerous and aggressive animals. Successful pet evacuation is a function of human and animal planners being at the same table prior to, during, and between disasters.

In most cases, local animal control has jurisdictional authority for pets in nonemergency times and in many cases will be designated as the local authority for pets during an evacuation or emergency. At the state level, typically, the state Department of Agriculture, State Veterinarian, or the Board of Animal Health support communities if they are overwhelmed. At the federal level, FEMA has the overall responsibility for supporting the states with pet issues, and the USDA Animal and Plant Health Inspection Service (APHIS) Animal Care provides subject matter expertise.

DEVELOP AN ANIMAL EVACUATION PLANNING COMMITTEE

When developing an animal evacuation planning committee, include representatives from animal control, animal welfare, veterinary medicine, transportation specialists, emergency services, and emergency management.

The first step is to estimate the number of household pets that may need assistance in evacuating, as well as the number of household pets within the community and the number of those household pets that may need to be evacuated and/or sheltered.

Analyze the human population and determine the following:

- The number of people and pets expected to evacuate on their own without assistance from local state or federal government
- The number of people and pets that will need assistance with sheltering
- The number of people and pets that will need evacuation and transportation assistance and sheltering

Historical data from previous evacuations and disasters can be used to determine the number of pets needing assistance in an evacuation. However, since enabling people with pets to evacuate is a relatively new concept and difficult to track, there are limited data. Therefore, the ideal method to estimate the number of animals to be evacuated is the preregistration of that population.

---

[5] Louisiana State Animal Response Team, 2008.

[6] https://www.avma.org/News/JAVMANews/Pages/131201a.aspx.

In Louisiana, animal planners have used the American Veterinary Medical Association's animal ownership formulas[7] to estimate the number of pets associated with a population of people. Animal planners used this formula to plan for a coastal evacuation for Hurricane Gustav. This formula overestimated the number of pets actually requiring evacuation and sheltering assistance significantly. A little over 37,000 people were identified as needing evacuation assistance for Hurricane Gustav, and 12,000 animals were expected to be associated with this population of people and also need assistance. Only 1200 (03.2%) pets actually accompanied their owners and required evacuation and sheltering assistance.[8]

The next step is to identify locations of animal businesses such as animal shelters, humane organizations, veterinary offices, boarding kennels, breeders, grooming facilities, human hospitals, nursing hospitals, assisted-living facilities, schools, animal testing facilities, or other animal-related entities. Emergency management may need to work with these facilities on developing their evacuation plans with specific attention to animal populations.

These entities should be identified as to name of facility, mailing address, physical address, owner's name and mailing address, manager's name, facility phone and fax as well as other emergency numbers, types of facility, types of animals at facility, how the animals are identified, how the animals are transported out of emergency situations, destination of evacuated animals, and how they are sheltered. Requesting this type of information annually encourages these animal-related entities to formulate and maintain up-to-date evacuation plans. Additionally, those in leadership positions have updated information to answer questions if the need arises.

IDENTIFY LOCATIONS THAT CAN BE USED FOR EMBARKATION/ COLLECTION POINTS

It's critical that embarkation points be identified well before a disaster and that information be distributed throughout the community. Embarkation points are gathering places for people and pets to be picked up and taken either to local receiving shelters, if the evacuation is from a localized event, or to a designated collection point or embarkation point to be registered and transported out of the area and possibly across county or state lines if the event is regional in scope.

Prior to identifying an embarkation point, emergency planners must take into consideration adequate parking, traffic flow, easy access for large vehicles, convenient area for loading vehicles, weather-protected areas, security, and staffing including a veterinarian.

IDENTIFY LOCATIONS FOR CO-LOCATED AND COHABITATED SHELTERS

Identify multiple facilities within each jurisdiction that can be used as pet shelters. These facilities should be identified and, depending upon the nature of the emergency, publicized so people know there is a place for their pets if they evacuate. For example, in San Diego County, the local animal control has numerous preapproved sheltering locations but does not announce the locations until the threat has been assessed.

DEVELOP PET REGISTRATION PROGRAM

Develop a pet registration system for pets that is integrated with the human registration. Whenever people are separated from their pets, a system of identifying the pet as owned and linking the pet to the owner should be used to insure that the pet is reunited with the owner. The pet registration section, if possible, can be set up in conjunction with human registration. In many cases, jurisdictions have found it beneficial to register pets first as they come into the registration center. After pets are registered, pet owners are then registered through human registration.

Some jurisdictions have found it useful to separate pet and nonpet owners at registration to avoid conflicts, which will decrease the chance of pet allergy problems in nonpet owners and will facilitate transportation to colocated pet shelters.

---

[7] https://www.avma.org/KB/Resources/Statistics/Pages/Market-research-statistics-US-pet-ownership.aspx.
[8] LSART, 2008.

PET EVACUATION PROCESS

In planning for the evacuation process, consider the following:

- Adequate resources/supplies are prestaged at embarkation point, including a supply of pet carriers for owners who do not bring their pets in a carrier.
- A veterinarian at the embarkation or collection point.
- Identify a local veterinary facility willing to accept pets if they become ill at the embarkation point.
- Identify veterinarians along the evacuation route willing to accept pets if they need medical attention during evacuation.
- A veterinarian at the receiving shelter when animals arrive.
- Educate and prepare first responders to assist people who own pets.

ACCEPTABLE TRANSPORTATION METHODS

Determine the method of acceptable transportation for the entire evacuation process, which may include short and long hauls. Mass transit vehicles used to transport people can effectively be used to transport people and their household pets resulting in less chance for people to become separated from their pets. Following Hurricane Sandy (2012), the City of New York opened up all public modes of travel and human evacuation shelters to pets.[9] And effective as of October 2018, New York state will allow people to take pets on **any kind** of public transportation when evacuating during an emergency.

Make sure the human transportation agreements specify that pets can accompany their owners on the vehicle during evacuation. Ensure that your evacuation plan includes a return plan that ensures that all pet owners are reunited with their pets and returned to their homes or provided shelter within the jurisdiction until reunification can occur.

SPECIAL PLANNING CONSIDERATIONS

When preparing an evacuation/transportation plan, consider the following:

- Multiple pet households
- Animal-related business such as boarding facilities and clinics that may require assistance
- Service animals
- Vaccination requirements
- Aggressive animals
- Human bite cases
- Nontraditional and exotic pets
- Non-English-speaking households
- Special needs owners
- Abandoned pets
- Pets with medical special needs
- Veterinary medical emergencies during evacuations
- Deaths during evacuations

MODES OF TRANSPORTATION

Many potential modes of transportation exist for evacuation of animals, ranging from the cars of private citizens to specially outfitted trucks. The vehicles most commonly used in large-scale transportation operations are the following:

- Transport refrigeration units
- Purpose-built animal transporters
- Modified animal welfare units

[9] http://www.cnn.com/2012/10/31/living/pets-superstorm-sandy/index.html.

- Animal control units
- Public transportation
- Farm/agriculture vehicles
- Box vans
- Climate-controlled cargo vans
- Personal occupied vehicles

No mode of transportation guarantees animal health and safety and even the best transport vehicle will only be as good as its operators. The following strengths and weaknesses are based on collective experience and consultation with subject matter experts, but weather, time, animal species/breed, terrain, airflow, accessibility, and vehicle operation, among other variables, can affect vehicle effectiveness.

The single most important issue facing transporters is the environmental conditions during transport. Adequate airflow, temperature control, ability to monitor air quality, and the presence of redundant systems were just a few of the features analyzed when reviewing the different types of vehicles for long-range transport. The absence of features for environmental control in a vehicle does not preclude its use, especially for short distances or in an emergency; every vehicle listed in this document has been used for emergency transport successfully in recent years by taking precautions to ensure that ventilation and airflow are maintained.

In summary, the single best practice for emergency planners during transportation is constant monitoring: when in doubt, check it out. In most cases, it is far better to add an hour to the trip than to lose an animal.

TRANSPORT REFRIGERATION UNITS

Commonly referred to as "reefers," transport refrigeration units (TRUs) are commonly used to haul perishable freight at specific temperatures. A refrigeration unit at the front of the trailer circulates the air in the trailer. Cooling units are designed to maintain the entire cargo area at constant temperature. Trailers are available to cool at a wide range of temperatures: from slightly cool for transporting items like produce (and animals) to freezer units for keeping items frozen through transport.

Refrigerated trailers range from 28 to 53 ft in length and from 96 to 102 in. in width. The most common units are 48–53′ long. TRUs have been used extensively over the years for short hauls of animals. Current practice is to stack two rows of large crates along each side of the trailer and to stop every 2 h to ventilate the trailer for 30 min, as animals produce gases, moisture, and heat. Most TRUs have front and rear ventilation doors that will assist in bringing in fresh air and removing "bad" air, but they are not effective in maintaining ventilation with large loads. Typically, it is more expensive to haul animals than produce as the cooling and ventilation systems work overtime to deal with respired gases (Fig. 8.1).

FIG. 8.1   Transport refrigeration unit (TRU). *Courtesy of Wikimedia Commons/D'oh Boy (Mark Holloway).*

Advantages

1. Readily available from commercial sources.
2. Ability to move large numbers of animals, primarily dogs and cats (average 90–100 dogs).
3. Cost-effective for large numbers of animals. Assuming 100 dogs/load and $1.73 per mile,[10] the average cost per animal per mile for refrigerated units is $0.017.
4. Multipurpose and easy to configure.
5. Ability to monitor temperature inside trailer from tractor.
6. Effective for large numbers of animals in hot weather.

Disadvantages

1. Not suitable for birds, rabbits, or pets with special needs that require consistent monitoring.
2. Under current USDA recommendations, the vehicle can only load two stacks of crates and can only operate for 2 h before having to stop for 30 min, ventilate the trailer, and check animals.
3. Extended time to fill trailer with animals may require the closing of one or both doors to moderate temperature.
4. May have inadequate airflow.
5. When idling, two motors will be running, decreasing efficiency and increasing carbon output.
6. Inability to monitor air quality in trailer from within cab.
7. Not cost-effective for small loads.
8. Requires a commercial driver's license to operate.
9. Can be difficult to safely secure crates without the correct equipment and/or tie-downs.
10. May be difficult to clean for trailers with wooden floors.
11. No direct access from cab to trailer.
12. May not be able to access remote areas due to height and length restrictions.
13. Will require loading ramps or docks or multiple people to lift crates due to height of trailer.
14. May not have communication capabilities that are interoperable with emergency services.
15. Heated units not readily available.
16. Noise from idling motors and cooling system may not be suitable for some neighborhoods.
17. Professional transport companies may not have any experience handling animals or be able to inspect animals while en route.

Recommendations

1. Request trailers with aluminum floors for easier cleaning/sanitizing.
2. Be aware of potential need for load bars, straps, etc.
3. Preplanning is required at local level for numbers of animals to ensure cost-effectiveness.
4. Secure a transportation management contract prior to incident.
5. In cold climates, provide bedding/blankets to keep animals warm.
6. Ventilate trailer and check animals every 2 h—more frequently if transporting special needs pets.
7. In warm climates, close doors every 15 min to stabilize temperature in the trailer.
8. Have safety officer perform safety check of vehicles before and during operation.
9. Have contact information for emergency repair service in case vehicle breaks down.
10. Have contact information for veterinary services along projected route.
11. Establish two- or three-deep contact information between drivers and senders and receivers.
12. Provide emergency food and water supplies.
13. Have safety officer ensure that crates are secured properly before and during trip.
14. When possible, identify loading and off-loading sites that offer loading docks suitable for height of trailer or sites where ramps will be usable.

---

[10] http://www.truckinfo.net/trucking/stats.htm.

15. When possible, provide handheld radio for driver to communicate with emergency services.
16. Have a dedicated team assigned to oversee animal loading, care, and inventory.
17. Have transporter provide specifics on vehicles including size, configuration, capacity, refrigeration settings, and exterior lighting.
18. Check ingress and egress routing and bridge heights to ensure vehicles will be able to access intake and export sites.

PURPOSE-BUILT ANIMAL TRANSPORT VEHICLES

These vehicles are custom-built for the transportation of animals and currently are the best method of animal transport available (Fig. 8.2). National nongovernmental, local/state animal rescue, and for-profit groups are putting custom-built units in service. Their biggest selling feature is the ability to maintain a constant temperature while providing the ventilation needed to avoid a buildup of toxic gases. Given their relatively large hauling capacity and controlled climate environment, they are the preferred choice of transport. Operating costs per mile may be slightly less to the TRUs as they have a smaller profile and tractors are generally more fuel-efficient.

Ventilation is through intake fans at the front of the trailer (usually on the ceiling) and exhaust fans in the floor area at the back of the trailer. These trailers can range in size from 20 to 53 ft and typically are 8' in width. Many of these custom-built units use a gooseneck hitch system that allows the tractors to be used for other purposes when disconnected.

Advantages

1. They are becoming more available.
2. Better ventilation than refrigerated trailers, provided vehicle has intake and exhaust fans.
3. Some of the units provide monitors and alarms to advise the driver if temperature, CO, CO_2, or O_2 reach dangerous levels.
4. Relatively large numbers of animals (60–100) can be moved. The TRUs may have the advantage in terms of capacity if triple stacked.
5. In most cases, it is easier and safer to load animals. Purpose-built vehicles typically have lower loading heights and drop-down rear doors that can be used as ramps.
6. Tractor can be used as a rescue vehicle when not towing trailer.
7. May be cost-effective for larger loads (10–15 mpg and 400 mi range).
8. Typically have side doors that provide easier access and temperature control.
9. Typically, these units are staffed with experienced animal welfare workers that will be able to assist in the loading and care of animals while being transported.

FIG. 8.2   ASPCA purpose-built transport vehicle. *Courtesy of The American Society for the Prevention of Cruelty to Animals. All rights reserved.*

Disadvantages

1. Not as available as TRUs.
2. Most cases will require a commercial driver's license (CDL).
3. Most units do not have interoperability (radio communication) with local emergency services.
4. Steps must be taken to ensure adequate air exchange within trailer and that monitoring systems are working properly.
5. Due to design considerations, initial cost to "build" is high, ranging upward from $200,000.
6. High cost precludes most local jurisdictions from owning.
7. Requires greater maintenance than refrigerated trailer due to more systems on board.
8. In cold winter climates, water lines may need to be drained to prevent freezing.
9. There is less ground clearance when compared with TRUs and other units, and care must be taken to avoid situations where the trailer may "belly up." Many of the custom-built units also have small living quarters and exposed plumbing under the trailer—adding to clearance challenges.

Recommendations

1. Operators of vehicles exceeding 26,000 lb gross vehicle weight rating (GVWR) must have a current commercial driver's license.
2. Provide transport groups with handheld radio so that they can communicate directly with emergency services.
3. Establish two- or three-deep contact information between drivers and senders and receivers.
4. Have transporter provide specifics on vehicles including size, configuration, capacity, built-in cages, generators, monitoring capability, water, exterior lighting, and temperature control.
5. Be aware of potential need for load bars, straps, etc.
6. Preplanning is required at local level for numbers of animals to ensure cost-effectiveness.
7. Secure a transportation management contract prior to incident.
8. Have safety officer perform safety check of vehicles before and during operation.
9. Have contact information for emergency repair service in case vehicle breaks down.
10. Have contact information for veterinary services along projected route.
11. Provide emergency food and water supplies.
12. Have safety officer ensure that crates are secured properly before and during trip.
13. When possible, identify loading and off-loading sites that offer loading docks suitable for ramp-style loading.
14. Have a dedicated team assigned to oversee animal loading, care, and inventory.
15. Check ingress and egress routing and bridge heights to ensure vehicles will be able to access intake and export sites.

MODIFIED ANIMAL WELFARE UNITS

These vehicles are typically operated by local (CARTS) (Fig. 8.3), state (SART), and/or national animal welfare groups. The trailer in the photo above is a multipurpose rig used for a SART program and has been used for adoption fairs, sterilization clinic, outreach projects, equipment and animal transport, and housing for rescue workers. The unit is self-contained and has radio and satellite communication systems. The back end is open and can transport 30–40

FIG. 8.3 Louisiana State Animal Response Team transport trailer. *Courtesy of LSART.*

animals depending on configuration. The advantage to having an open cargo bay is that this same unit has hauled birds from an oil spill, domestic birds from a seizure case, stranded marine mammals, and even three tigers. Most of these vehicles come with multiple air conditioning units, heater, generator, interior and exterior lighting, drop-down ramp, and side-entry doors. They can be pulled easily with 1T vehicle and generally utilize a gooseneck hitch system.

Advantages

1. Becoming more available.
2. Vehicles typically come with personnel experienced in animal handling.
3. Some vehicles may have veterinary facilities/supplies/equipment onboard.
4. Tractor (pull vehicle) can used as a tow/rescue vehicle when not towing trailer.
5. Generally well ventilated and temperature controlled.
6. Generally easier to access difficult areas, and two vehicles are often 4WD.
7. Generally low profile and can access areas with overhead restrictions.
8. Generally multipurpose and, as such, easier to configure cargo bay for different sized species and/or crates.
9. Generally more cost-effective to operate than TRUs and custom-built units (12–15 mpg and 350–400 mi range).
10. Depending on configuration, may be able to transport 30–40 animals.
11. In most cases, it is easier and safer to load animals versus TRUs. Modified animal welfare units typically have lower loading heights and drop-down rear doors that can be used as ramps.
12. Typically have side doors that provide easier access and temperature control.
13. Generally less expensive to construct and operate than custom-built units.
14. Tow vehicles typically have diesel engines that are more economical and provide greater torque than gasoline engines.

Disadvantages

1. Not readily available.
2. May be designated to serve as mobile veterinary hospital, thus precluding use as a transport vehicle.
3. May not have interoperability (radio communication) with local emergency services.
4. Expensive—upward of $150,000.
5. Depending on the GVWR, may require a commercial driver's license (CDL).
6. Steps must be taken to ensure adequate air exchange within trailer and that monitoring systems are working properly.
7. Requires greater maintenance than refrigerated trailer due to more systems on board.
8. In cold winter climates, water lines may need to be drained to prevent freezing.
9. There is less ground clearance when compared with TRUs and other units and care must be taken to avoid situations where the trailer may "belly up." Many of the modified animal welfare units also have small living quarters and exposed plumbing under the trailer—adding to clearance challenges.
10. Diesel fuel may be difficult to find in some communities.

Recommendations

1. Operators of vehicles exceeding 26,000 lb GVWR must have a current commercial driver's license.
2. Determine availability of vehicles from other agencies/jurisdictions prior to incident.
3. Provide transport groups with handheld radio so that they can communicate directly with emergency services.
4. Establish two- or three-deep contact information between drivers and senders and receivers.
5. Have transporter provide specifics on vehicles including size, configuration, capacity, built-in cages, generators, monitoring capability, water, exterior lighting, and temperature control.
6. Be aware of potential need for load bars, straps, etc.
7. Preplanning is required at local level for numbers of animals to ensure cost-effectiveness.
8. Secure a transportation management contract prior to incident.
9. Have safety officer perform safety check of vehicles before and during operation.
10. Have contact information for emergency repair service in case vehicle breaks down.
11. Have contact information for veterinary services along projected route.

12. Provide emergency food and water supplies.
13. Have safety officer ensure that crates are secured properly before and during trip.
14. When possible, identify loading and off-loading sites that offer loading docks suitable for ramp-style loading.
15. Have a dedicated team assigned to oversee animal loading, care, and inventory.
16. Check ingress and egress routing and bridge heights to ensure vehicles will be able to access intake and export sites.

ANIMAL CONTROL UNIT

Animal control units such as the one shown here (Fig. 8.4) from San Diego Co. Animal Services come in a variety of styles and sizes, ranging from fully outfitted vehicles to vans with either built-in or temporary caging. For the purpose of this document, pickups with or without a canopy and caging in the bed are not considered "animal control units."

Capacity for these units will range from four to eight animals, and storage bays will be ventilated and may come with climate control. The vehicles are typically on a 1/2T frame but may be available in 3/4T and 4WD. Many of these vehicles are outfitted with emergency lights, sirens, and communication equipment. In many cases, they are part of local government and thereby highly recognizable as an emergency vehicle. Some communities have typed and registered their vehicles and staff with State Emergency Management and may be available for an EMAC request.

Advantages

1. Usually available locally.
2. Immediately accessible.
3. May be available through mutual aid or intergovernmental agreements.
4. Recognized within jurisdiction and easily seen as an emergency vehicle outside jurisdiction.
5. Typically equipped with emergency lights and siren.
6. Normally come with an Animal Control Officer and animal control equipment.
7. Ideal for small numbers of animals and short-distance trips.
8. For small numbers, more economical than trailered units (15–20 mpg and 400 mi range).
9. Animal cages may be climate-controlled.
10. Most units have communication capabilities.
11. Cost is often borne by local jurisdiction. Units typically cost between $35 and $60K depending on equipment and chassis.
12. May be able to pull a trailer.
13. Better accessibility than trailered units and typically not restricted on any roadways.

Disadvantages

1. Can only transport a small number of animals. For small animals, multiple crates can be stored in each bay.
2. May not be climate-controlled and inappropriate for extreme weather and long hauls.

FIG. 8.4 Animal Control Unit. *Courtesy of San Diego County Animal Services.*

3. Typically cannot monitor conditions or animals in the animal cages. Some units may come equipped with sound monitoring.
4. May not be available in affected community due to involvement in the response efforts.
5. Communications equipment will have designated frequency that might not be compatible outside their jurisdiction.
6. In most cases, animals will need to be picked up and loaded into bays that may be challenging with large and/or fractious animals. Some vehicles will come with loading ramps.

Recommendations

1. Establish mutual aid agreements in advance for provision of vehicles and personnel.
2. Determine availability of vehicles from other agencies/jurisdictions prior to incident.
3. Provide transport groups with handheld radio so that they can communicate directly with emergency services.
4. Establish two- or three-deep contact information between drivers and senders and receivers.
5. Have transporter provide specifics on vehicles including size, configuration, capacity, monitoring capability, exterior lighting, and temperature control.
6. Preplanning is required at local level for numbers of animals to ensure cost-effectiveness.
7. Have safety officer perform safety check of vehicles before and during operation.
8. Have contact information for emergency repair service in case vehicle breaks down.
9. Have contact information for veterinary services along projected route.
10. Provide emergency food and water supplies.
11. Have a dedicated team assigned to oversee animal loading, care, and inventory. Due to the height of the bays, care should be taken when loading and off-loading animals.

PUBLIC TRANSPORTATION

Public transportation includes a wide range of vehicles including local school and metropolitan buses, long-distance buses, planes, and trains. A number of jurisdictions list these modes of transport in their emergency plans as they are generally easy to access, can hold large numbers of animals, and may allow animals to travel with their owners. The primary considerations for determining which of these modes would best fit a community's needs are configuration and loading. Public transportation is designed to hold large numbers of seated humans and is not generally configured in a way to handle crated animals. Doorways may provide limited access, and aisle ways may offer challenges for moving large crates. Bench seats typically are not conducive for strapping down crates.

Some jurisdictions allow animals to board a vehicle with their owner that works reasonably well as long as animals are tightly reined in or on the owner's lap. It is not recommended to load animals in commercial vehicles tethered by their leash. Unexpected movement can cause the dog to be thrown off the bench seat and hung up by its leash.

Airplanes are typically not cost-effective but can move a reasonably large number of animals in the shortest period of time. Following Hurricane Katrina, approximately 125 animals could be transported on commercial airframes (727, 737, and MD 80). Though expensive to operate and labor-intensive to coordinate and load, the short flight time can reduce a cross-country trip by 40–50h. Or in the more recent case of Hurricane Maria, there were no other acceptable modes of travel off the islands. Well, over 2500 animals were transported by plane off Puerto Rico, St. Thomas, and St. Croix. Following Hurricane Irma (2017), Miami Dade Animal Services, FedEx, and ASPCA coordinated the transport of well over 100 animals from Miami to rescue groups on the west coast using a 747 (Fig. 8.5). In most cases (the exception being some school buses), the environment within the "vehicle" can be controlled, and ventilation (air exchange) is adequate for transporting animals.

Advantages

1. Readily available in most jurisdictions depending upon transportation type.
2. Recognizable in the jurisdiction.
3. May be climate-controlled.
4. Convenient for people and animals to travel together.
5. Can move large numbers of animals.
6. Easy to monitor climate and animals.

FIG. 8.5 ASPCA and Miami-Dade Transport using 747 following H. Irma. *Courtesy of The American Society for the Prevention of Cruelty to Animals. All rights reserved.*

7. Comes with an experienced driver/pilot.
8. May have a local communications system.

Disadvantages

1. Due to the priority of evacuating people, public transportation may not be available.
2. School buses may not be climate-controlled and thus not suitable for hot climates and/or long hauls.
3. Difficult to load large crates through doorways of most buses and airplanes.
4. Difficult to secure animal crates in seats.
5. Concerns over allergies may preclude animals being transported with people.
6. Vehicles may be privately owned and not cost-effective (commercial buses, airplanes, and trains).
7. Planes and trains are limited to rail routes or flight paths and runways that will require another mode of transportation for loading and off-loading.
8. Requires additional planning/logistics.
9. May be cost-ineffective (planes and trains).
10. School buses are typically not suitable for long hauls (range and seat construction).
11. Buses may be restricted in access due to their length and/or height.
12. Diesel fuel may be difficult to find in some communities.

Recommendations

1. Investigate local availability of transportation types and any limitations pertaining to the transport of animals prior to incident.
2. Working group recommends that only licensed, qualified drivers operate vehicles.
3. Establish two- or three-deep contact information between drivers and senders and receivers.
4. Provide transport groups with handheld radio so that they can communicate directly with emergency services.
5. Have transporter provide specifics on vehicles including size, configuration, capacity, water, exterior/interior lighting, and temperature control.
6. Be aware of potential need for crates, load bars, straps, etc.
7. Preplanning is required at local level for numbers of animals to ensure cost-effectiveness.
8. Secure a transportation management contract prior to incident.
9. Have safety officer perform safety check of vehicles before and during operation.
10. Have contact information for emergency repair service in case vehicle breaks down.
11. Have contact information for veterinary services along projected route.
12. Provide emergency food and water supplies.
13. Have safety officer ensure that crates are secured properly before and during trip.

14. When possible, identify loading and off-loading sites that are conducive for mode of transportation.
15. Have a dedicated team assigned to oversee animal loading, care, and inventory.
16. Check ingress and egress routing and bridge heights to ensure vehicles will be able to access intake and export sites (buses).

FARM/LIVESTOCK VEHICLES

Farm/livestock vehicles (Fig. 8.6) are used to transport various types of livestock and can range from anything as small as a one-horse trailer to large livestock haulers. Commercial cattle transporters range from 48–53 ft in length by 8.5 ft in width. Other livestock (hogs and pigs) may be transported in 7 × 32–40 ft gooseneck trailers. Typically, beds are open and have similar capacity to TRUs. They are not climate-controlled though some come with sprinkler systems for summer transport. Ventilation is generally adequate, and as long as the vehicle is moving, conditions in the back are tolerable in summer. Temperatures can quickly rise in the summer if the vehicle is stopped. Most often rear-loaded using a ramp/door system, these trailers are easier to load than the TRUs. Ground clearance and height and length issues may limit access to some areas.

Advantages

1. Readily and immediately available.
2. Cost-effective for transporting large numbers of animals (5–7 mpg).
3. May be used for a variety of species.
4. Horse and small livestock trailers will have better ground clearance than other trailers.
5. May come with an operator experienced with livestock and other animals.
6. Most effective in temperate settings (60–70°C).
7. Easy to load and generally easy to secure crates through tie-downs and side slots.
8. Commercial haulers can hold approximately 100 large crates if double stacked.

Disadvantages

1. Maintenance records on vehicles may be unavailable.
2. May need to be cleaned/sanitized prior to use during incident.
3. Will likely not have any climate control capabilities.
4. May lack protection from the elements.
5. Possible traumatic ride conditions could cause stress on uncrated animals.
6. May require a special vehicle to tow the trailer.
7. Requires an experienced CDL driver.

FIG. 8.6 Livestock transport trailer. *Courtesy of Pixabay/Rene Rauschenberger.*

8. Do not have communications capabilities.
9. Unable to monitor conditions of the animals while driving.

Recommendations

1. Investigate local availability of transportation types and any limitations pertaining to the transport of animals prior to incident.
2. Only licensed, qualified drivers operate vehicles.
3. Provide transport groups with handheld radio so that they can communicate directly with emergency services.
4. Establish two- or three-deep contact information between drivers and senders and receivers.
5. Have transporter provide specifics on vehicles including size, configuration, capacity, and exterior/interior lighting.
6. Be aware of potential need for crates, load bars, straps, etc.
7. Preplanning is required at local level for numbers of animals to ensure cost-effectiveness.
8. Secure a transportation management contract prior to incident.
9. Have safety officer perform safety check of vehicles before and during operation.
10. Have contact information for emergency repair service in case vehicle breaks down.
11. Have contact information for veterinary services along projected route.
12. Provide emergency food and water supplies.
13. Have safety officer ensure that crates are secured properly before and during trip.
14. When possible, identify loading and off-loading sites that are conducive for ramp off-loading.
15. Have a dedicated team assigned to oversee animal loading, care, and inventory.
16. Check ingress and egress routing and bridge heights to ensure vehicles will be able to access intake and export sites.
17. When in doubt, thoroughly clean inside bay before hauling animals.
18. Be aware of weather forecast for projected route.

BOX VANS

Box vans (Fig. 8.7) are typical rental vans commonly used when people are moving from one residence to another. They come in a variety of sizes and capacities—with most common lengths ranging from 14 to 26 ft. The interior cargo bay height typically runs between 7 and 8 ft. They have been used in disaster response as a last line of defense for short hauls. There is little to no ventilation in the cargo bay, so doors must be opened frequently, and transporter needs to shut down the motor when doors are open to avoid exhaust fumes entering bay. Some box vans have a driver pass-through door that allows for limited air circulation in the cargo area. This mode of transportation may be safe for small species and small loads over short time frames.

Advantages

1. Usually readily available (depending upon season and geographic area).
2. Easy to drive and do not require a special license to drive, unless it is a large commercial vehicle.

FIG. 8.7 ASPCA Box van. *Courtesy of The American Society for the Prevention of Cruelty to Animals. All rights reserved.*

3. Some come with ramps for easier loading.
4. Generally cost-effective (10 mpg and range of 400 mi for 17 ft model).

Disadvantages

1. May have restrictions for use to transport animals.
2. Temperature/climate control varies by vehicle type.
3. Inability to monitor temperature and animals if no direct access from cab.
4. May lack adequate tie-downs.
5. Do not have communications capabilities.
6. Some have wooden floors, which makes it difficult to clean/sanitize.
7. No ventilation doors.

Recommendations

1. Operator must stop regularly to monitor animals and ensure adequate air temperature and quality.
2. Investigate local availability of transportation types and any limitations pertaining to the transport of animals prior to incident.
3. Even though box vans are relatively easy to drive, larger vans require a degree of driver training and expertise.
4. Provide transport groups with handheld radio so that they can communicate directly with emergency services.
5. Establish two- or three-deep contact information between drivers and senders and receivers.
6. Be aware of potential need for crates, load bars, straps, etc.
7. Preplanning is required at local level for numbers of animals to ensure cost-effectiveness.
8. Have safety officer perform safety check of vehicles before and during operation.
9. Have contact information for emergency repair service in case vehicle breaks down.
10. Have contact information for veterinary services along projected route.
11. Provide emergency food and water supplies.
12. Have safety officer ensure that crates are secured properly before and during trip.
13. When possible, identify loading and off-loading sites that are conducive for ramp and dock off-loading.
14. Have a dedicated team assigned to oversee animal loading, care, and inventory.
15. When in doubt, thoroughly clean inside bay before hauling animals.

CLIMATE CONTROLLED CARGO VANS

These vehicles are standard cargo vans or multipassenger vans (Fig. 8.8) with the seating removed. Size will vary depending upon the amount of cargo space or passenger room. Most common dimensions for inside cargo area are 108" × 67" × 53" (L × W × H). It is the intent that the animals are within the climate-controlled portion of the van. This mode of transportation has been used extensively for emergency transportation as they are readily available, easy to drive, and easy to control the air conditions in the cargo area. They are ideal for smaller species/crates but not ideal for double stacking, which reduces airflow. The vans shown below are modified vans used by the ASPCA Relocation Team and similar to the type of vehicle used following the Deepwater Horizon oil spill for transporting birds. In that case, the number of animals needing transport was small, and medium to large crates were used making the cargo van the preferred, cost-effective vehicle.

Advantages

1. Readily available in most, larger communities.
2. Easy to drive and do not require any special license.
3. Driver/passenger can easily monitor temperature and animals.
4. Ideal for small loads and typically "comfortable" for longer hauls.
5. Cost-effective for small loads (16 mpg and 500 mi range).
6. Easier to access most areas and generally not affected by height and length restrictions.

FIG. 8.8   ASPCA transport vans. *Courtesy of The American Society for the Prevention of Cruelty to Animals. All rights reserved.*

Disadvantages

1. Can only carry a limited number of animals.
2. Driver exposed to animals.
3. Do not have communications capabilities.
4. May have to remove seats to increase cargo size.

Recommendations

1. Recommended for transport of animals with special needs that need to be monitored regularly.
2. Investigate local availability of transportation types and any limitations pertaining to the transport of animals prior to incident.
3. Even though cargo vans are relatively easy to drive, larger vans require a degree of driver training and expertise.
4. Provide transport groups with handheld radio so that they can communicate directly with emergency services.
5. Establish two- or three-deep contact information between drivers and senders and receivers.
6. Be aware of potential need for crates, load bars, straps, etc.
7. Preplanning is required at local level for numbers of animals to ensure cost-effectiveness.
8. Have safety officer perform safety check of vehicles before and during operation.
9. Have contact information for emergency repair service in case vehicle breaks down.
10. Have contact information for veterinary services along projected route.
11. Provide emergency food and water supplies.
12. Have safety officer ensure that crates are secured properly before and during trip.
13. Have a dedicated team assigned to oversee animal loading, care, and inventory.
14. When in doubt, thoroughly clean inside bay before hauling animals.

PERSONALLY OPERATED VEHICLES

These vehicles are personal cars or trucks (Fig. 8.9) called in to transport animals. The intent is that the animal is transported within the climate-controlled cabin of the vehicle. Typically used as a last resort given the limited capacity and inability to separate animals from driver. Since most personal vehicles have a bench-style back seat, it is difficult to load more than two medium to large crates.

Advantages

1. Readily available and usually comes with a driver.
2. Easy to drive.
3. Usually climate-controlled.
4. Easy to monitor animals.

FIG. 8.9   Personl Operated Vehicle (POV). *Courtesy of Wikimedia Commons/Kjell Lundberg.*

5. No special license required beyond driver's license.
6. Driver generally aware of maintenance records and vehicle limitations.
7. Easy access to most areas.
8. Generally fuel-efficient.

Disadvantages

1. May not be adequately insured.
2. Proof of ownership may be lacking.
3. May not be climate-controlled.
4. Inability to transport large numbers of animals.
5. Not cost-effective to transport large numbers of animals.
6. Drivers exposed to animals during transport.

Recommendations

1. Require proof of current driver's license, registration, and insurance prior to use.
2. Recommended for transport of a small number of animals with special needs that need to be monitored regularly.
3. Provide transport groups with handheld radio so that they can communicate directly with emergency services.
4. Establish two- or three-deep contact information between drivers and senders and receivers.
5. Have safety officer perform safety check of vehicles before and during operation.
6. Have contact information for emergency repair service in case vehicle breaks down.
7. Have contact information for veterinary services along projected route.
8. Provide emergency food and water supplies.

CHAPTER

9

Hazards

Different regions of the country face the potential for different types of disasters. The northwest is prone to floods and fires, the west coast sees large numbers of fires and earthquakes, the Midwest experiences the greatest number of tornadoes, the Gulf and Mid-Atlantic see the greatest number of hurricanes, and the northeast is known for its winter storms (Fig. 9.1).[1]

The United States is subjected to just about every type of severe weather event imaginable with floods being most frequent followed by wind events (storms), earthquakes, and extreme weather events (Fig. 9.2). If you would like to live in the safest place in the United States in terms of potential impact of disasters, you may want to consider Corvallis, OR, and/or if you are feeling on the bold side, you may want to move to just about anywhere in Texas.[2]

Flooding accounted for 47% of the total number of *relevant* events last year worldwide and 65% of the fatalities resulting from disasters (Fig. 9.3). Meteorologic events (tropical storm and thunderstorm) were responsible for the greatest loss of property.

HURRICANES

Other than some small differences in wind speed, there is no difference between a hurricane, typhoon, or cyclone. They are all different names for the same kind of intense low-pressure system. There are two different kinds of cyclones that affect the United States most often: tropical cyclones and extratropical cyclones.

A tropical cyclone describes a rotating, organized system of clouds and thunderstorms that originates over tropical or subtropical waters and has closed, low-level circulation.

The weakest tropical cyclones are called *tropical depressions*. If a depression intensifies to maximum sustained winds of 39 mph, it becomes a *tropical storm*. If a tropical storm reaches maximum sustained winds of 74 mph or higher, it becomes a hurricane, typhoon, or tropical cyclone, depending upon where the storm originates in the world. In the North Atlantic, central North Pacific, and eastern North Pacific, it's referred to as a *hurricane*. In the Northwest Pacific it's called a *typhoon*, and in the South Pacific and Indian Ocean, it's called a *tropical cyclone*.

The ingredients needed for tropical cyclone formation include a preexisting weather disturbance, warm oceans, moisture, and relatively light winds. If those conditions persist over an extended period of time, they can combine to produce the potentially violent and destructive storm systems all too prevalent for the coastal regions of the United States.

In the Atlantic, hurricane season officially runs from June 1 to November 30. Ninety-seven percent of tropical cyclone activity occurs during this time period (Fig. 9.4).

Hurricanes are categorized by wind speed using the Saffir-Simpson Hurricane Scale:

- *Category 1*: Winds 74–95 mph
- *Category 2*: Winds 96–110 mph
- *Category 3*: Winds 111–129 mph
- *Category 4*: Winds 130–156 mph
- *Category 5*: Winds more than 157 mph

[1] https://web.archive.org/web/20140717040901/http://www.crisishq.com/why-prepare/us-natural-disaster-map.

[2] https://www.bestplaces.net/docs/studies/avoid_natural_disasters.aspx.

FIG. 9.1 Maps depicting the number of times each state has been affected by each of six types of billion dollar weather and climate disasters. *Image courtesy of* NOAA/Climate.gov.

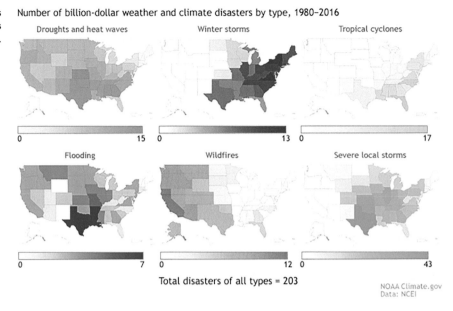

FIG. 9.2 Normalized Hazard Risk Score using nine commonly found hazards in the United States to predict probablity of loss. *Courtesy of CoreLogic, Inc.*

A hurricane has a number of identifiable parts including the following:

- *Eye*: The eye is the "hole" at the center of the storm. Winds are light, and skies are only partly cloudy, sometimes even clear, in this area.
- *Eye wall*: The eye wall is a ring of thunderstorms swirling around the eye. The wall is where winds are strongest and rain is heaviest.
- *Rain bands*: Spiral bands of clouds, rain, and thunderstorms extend out from a hurricane's eye wall. These bands stretch for hundreds of miles and sometimes contain tornadoes.

Hurricane development and the rate at which wind speed or direction changes with height is referred to as vertical wind shear. Low vertical wind shear—winds that change very little going up through the atmosphere—is needed for

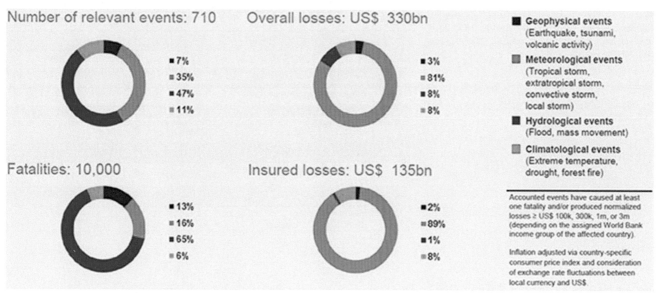

Number of relevant events: 710

- 7%
- 35%
- 47%
- 11%

Overall losses: US$ 330bn

- 3%
- 81%
- 8%
- 8%

■ **Geophysical events**
(Earthquake, tsunami, volcanic activity)

■ **Meteorological events**
(Tropical storm, extratropical storm, convective storm, local storm)

■ **Hydrological events**
(Flood, mass movement)

■ **Climatological events**
(Extreme temperature, drought, forest fire)

Accounted events have caused at least one fatality and/or produced normalized losses ≥ US$ 100k, 300k, 1m, or 3m (depending on the assigned World Bank income group of the affected country).

Inflation adjusted via country-specific consumer price index and consideration of exchange rate fluctuations between local currency and US$.

Fatalities: 10,000

- 13%
- 16%
- 65%
- 6%

Insured losses: US$ 135bn

- 2%
- 89%
- 1%
- 8%

FIG. 9.3 Impact of disasters on the loss of life and property losses worldwide. *Insurance Information Institute.*

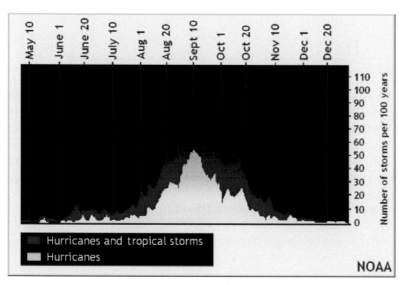

Hurricanes and tropical storms
Hurricanes

NOAA

FIG. 9.4 Number of tropical cyclones in the United States by month. *Courtesy of NOAA.*

hurricane development. High vertical wind shear—winds that are changing significantly with height—tends to rip storms apart.[3]

The National Hurricane Center (NHC) issues forecasts using the following categories:

Hurricane advisory tells where the storm is located, wind speeds, and direction.
Hurricane watch is issued when there is a threat of hurricane conditions within 24–36 h.
Hurricane warning is issued when hurricane conditions are expected in 24 h or less.

The United States has experienced a number of devastating hurricanes over the last two centuries. The 1900 Galveston Hurricane is on everyone's top 10 list with estimated deaths between 8 and 12,000 people. About 28 years later, an estimated 2500 people drowned when a Category 4 hurricane caused Lake Okeechobee in Florida to overflow, deluging the surrounding area with 10–15 ft floods. In terms of wind speed, Hurricane Michael (Oct. 2018) hit the Panhandle of Florida with 155 mph sustained winds making it the third strongest hurricane in US history. But Hurricane Patricia (Oct. 2015) holds the global record with 1-minute sustained winds of 215 mph! The three storms with the greatest animal impact over the last 25 years would be Hurricane Andrew (1992); Hurricane Floyd (1999);

[3] https://www.nasa.gov/audience/forstudents/5-8/features/nasa-knows/what-are-hurricanes-58.html.

FIG. 9.5 Ten deadliest mainland US tropical cyclones causing 25 or greater deaths, 1851–2010. *Table adapted from NOAA Technical Memorandum NWS NHC-6.*

Rank	Name	Year	Category	Deaths
1.	Great Galveston Hurricane (TX)	1900	4	8000
2.	FL (Lake Okeechobee)	1928	4	2500
3.	Katrina (LA/MS/FL/GA/AL)	2005	3	1200
4.	Cheniere Caminanda (LA)	1893	4	1100-1400
5.	Sea Islands (SC/GA)	1893	3	1000-2000
6.	GA/SC	1881	2	700
7.	Audrey (SW LA/N TX)	1957	4	416
8.	Great Labor Day Hurricane (FL Keys)	1935	5	408
9.	Last Island (LA)	1856	4	400
10.	Miami Hurricane (FL/MS/AL/Pensacola)	1926	4	372

and, of course, Hurricane Katrina (2005). The following is a list developed by Weather Underground of the top 10 deadliest hurricanes in US history (Fig. 9.5).[4]

FIRES

An average of 5 million acres burns every year in the United States. Western states experience some of the worst of the nation's forest fires. In 2017, California experienced the most destructive fire season on record with over 9000 fires and burning more than a million acres.[5] Two particularly devastating fires occurred late in the year with the Tubbs Fire in Sonoma County (Sep.) and the Thomas Fire in Ventura County (Dec.). Collectively, these fires resulted in the loss of over 10,000 structures, 44 deaths, and hundreds injured.[6] AccuWeather predicted that the total economic toll for the 2017 wildfire season would be at least $180 billion.[7] In August of 2018, the Mendocino Fire Complex became the largest fire in California state history burning over 400,000 acres. Lake County Animal Control and the ASPCA assisted over 5600 animals in a two week period of time.

Fires typically leave the soil more prone to movement (landslides), and potential flooding issues occurred in January 2018 in Santa Barbara County when massive landslides occurred after torrential rain fell on ground burned from the Thomas Fire. With no roots or foliage to help absorb the water, the ground let go and buried sections of Montecito, closing down Highway 101 and killing at least 21 people and injuring many more.

Fire Behavior

Three elements must be present for fire to start: fuel source, air, and heat. In the western states and specifically California, there are plenty of fuel sources available due to the excessive moisture early in 2017 and then the return to extended dry and hot weather. Consequently, what hasn't already burned is likely a viable fuel source for the next fire season. Interestingly, the traditional fire season is becoming longer as witnessed by the early (GA/FL) and late (CA) season fires in 2017. As the California governor recently stated, this is the "new normal."[8] Fires are becoming more expensive to manage, and in 2015, the United States broke $10 million in firefighting costs for the first time (Figs. 9.6 and 9.7).

[4] https://www.wunderground.com/hurricane/usdeadly.asp.

[5] Dale Kasler. "Wine country wildfire costs now top $9 billion, costliest in California history." The Sacramento Bee, December 8, 2017.

[6] "2017 Fire Statistics." CAL FIRE. Retrieved April 22, 2017.

[7] "AccuWeather predicts 2017 California wildfire season cost to rise to $180 billion." AccuWeather. December 8, 2017.

[8] http://www.latimes.com/local/lanow/la-me-socal-fires-20171210-story.html.

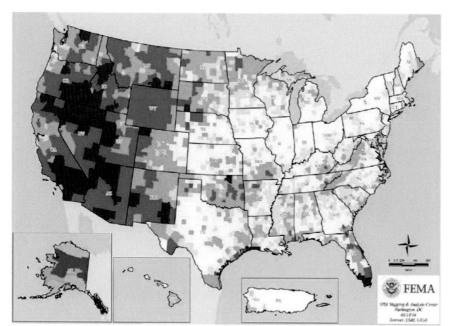

FIG. 9.6 Fire Activity by County. *Courtesy of the Federal Emergency Management Agency (FEMA)/Jana Baldwin.*

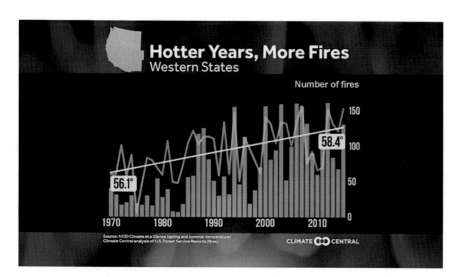

FIG. 9.7 Number of fires in western states from 1970–2016. *Courtesy of Climate Central. All rights reserved.*

FLOODING

Many experts believe that as global warming continues and we see a disproportionate cooling in the northern latitudes, we will see more polar vortices and severe weather events in the middle latitudes (the United States). In the last quarter of 2013 through January of 2014, we experienced record flooding in the Midwest, unusually early record snowfall in the Dakotas, tornadoes in the Midwest and East Coast, polar vortex that shattered records followed by temps in the 1970s in New York City, and significant fire danger and drought in California. The summer of 2015 saw record (whiplash) flooding throughout Texas and Oklahoma, and 2016 was the worst flood season on record for the United States.

A single weather event does not support or disprove global climate warming, but regardless of the cause, we are experiencing significant weather events, and there is nothing to suggest that is going to change. The impact of climate change on animals will be discussed in Chapter 13.

Much of the United States is impacted from flooding, and all 50 states have experienced flash flooding. Areas most prone to flooding are shown in Fig. 9.8.

FIG. 9.8 Frequency of flood events by county. *Courtesy of the Federal Emergency Management Agency (FEMA).*

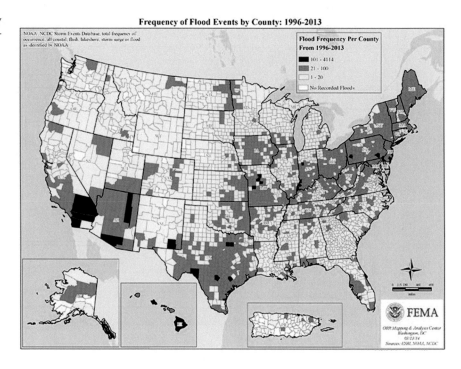

FIG. 9.9 Weather fatalities 2017. *Courtesy of NOAA National Weather Service.*

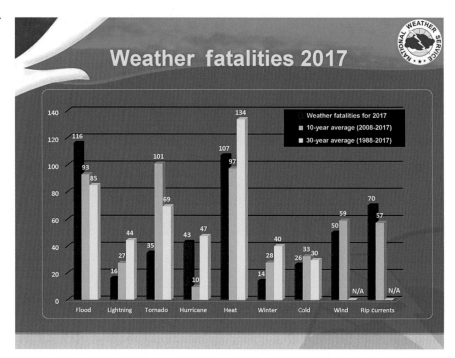

As mentioned previously, flooding is a major contributor to the loss of human life in the United States (Fig. 9.9). That is also true for animal life as evidenced in a number of recent floods (2016, 2017). Urban areas with their expansive network of asphalt and concrete are more prone to flooding, and the NE to Midwest regions where snowmelt combines with rain are prone to flooding on a near-annual basis. The Red River that separates Minnesota and North Dakota, for example, has flooded 49 of the last 110 years. Fortunately, flood-control efforts since the great flood of 1997 have lessened the flooding risk significantly.

VOLCANIC ERUPTIONS

There are about 1500 potentially active volcanoes worldwide,[9] and according to the US Geological Survey, there are 169 active volcanoes in the United States and its territories. Fifty four of those are deemed a *very high* or *high* threat to public safety.[10] Over half a million people live within 20 mi of a volcano.[11] The most destructive volcanic eruption in US history was Mount St. Helens on May 18, 1980. The blast killed 57 people and thousands of animals[12] leveling 200 sq mi of forest from a landslide that came down at speeds predicted to reach 300 mph. The mountain had been stirring for 2 months prior to exploding. A 5.1-magnitude earthquake triggered a sideways blast that ripped off the mountain's north face and sent a plume of ash 16 mi in the air, eventually covering three states. Scientists say that Mount St. Helens is "the most likely of the contiguous US volcanoes to erupt in the future."[13]

More recently, the Kilauea volcano on the Big Island of Hawaii has experienced volcanic activity and lava flow twice in recent years (2014 and 2018). In May of 2018, lava flow erupted from the volcano and destroyed more than 100 homes and forced the evacuation of thousands from the east side of the island. Hundreds of animals were evacuated with their owners, and an equal number were later rescued by the various animal welfare groups on the island. Kilauea is one of the most active volcanoes in the world and has been erupting on and off for hundreds of thousands of years. 6, 2018 near Pahoa, Hawaii (Fig. 9.10).

In June of 2018, a catastrophic eruption of Mt Fuego in Guatemala killed at least 75 people and devastated a number of villages that were in the path of the deadly ash and lava flow. Unlike Hawaii's Kilauea volcano, which mostly produces slow-moving lava flows, Fuego is a different type of volcano, known as a stratovolcano, generating multiple pyroclastic and ash flows. A pyroclastic flow is a fast-moving mixture of ash, volcanic rock, hot gases, and air, and it can reach temperatures in excess of 500 °C and will destroy anything in its path. When Mount St. Helens erupted in Washington State in 1980, it was a pyroclastic flow that did so much damage (Fig. 9.11).

The biggest dangers to animals from a volcanic eruption are lava and air quality. Lava can reach temperatures exceeding 2000 °C, and as it comes down the mountain, it often breaks apart into hot rock and gas (pyroclastic) flow. And as mentioned, this flow will destroy everything in its path including buildings, cars, and even bridges. In 1902, Mount Pelée located on the French Island of Martinique erupted killing at least 15,000 people and destroying the

FIG. 9.10 Fissure 7 photographed May 5, 2018 during the Kilauea volcanic eruption. *Image credit: Department of the Interior/USGS.*

[9] https://www.usatoday.com/story/news/nation/2018/05/07/hawaii-volcano-science-behind-eruption-kilauea/586268002/.

[10] https://volcanoes.usgs.gov/vsc/file_mngr/file-153/FAQs.pdf.

[11] https://www.preventionweb.net/countries/usa/data/.

[12] http://content.time.com/time/specials/packages/article/0,28804,2014572_2014574_2014629,00.html.

[13] https://www.nytimes.com/2018/05/14/us/us-active-volcanoes-hawaii.html.

FIG. 9.11   May 18, 1980 eruption of Mount St. Helens eruption, viewed from the south. *Image credit: Department of the Interior/USGS.*

entire town of St Pierre. Before lava enveloped the city, the earthquakes and eruptions caused an army of creatures to come out of hiding and seek safety in the city. As reported in *Earth*,[14] a plague of insects, snakes, gigantic centipedes, and deadly 2 m-long pit vipers came out of the forest and raced into the city claiming the lives of hundreds of livestock and about 50 people.

Magma is molten rock that is below the Earth's crust and contains dissolved gases, which provide the driving force that causes most volcanic eruptions. As magma rises toward the surface, gases are released from the liquid portion of the magma and continue to travel upward and are eventually released into the atmosphere. Large eruptions can release enormous amounts of gas in a short time. The most abundant volcanic gas is water vapor, which is harmless. However, significant amounts of carbon dioxide, sulfur dioxide, hydrogen sulfide, and hydrogen halides can also be emitted from volcanoes. Depending on their concentrations, these gases are all potentially hazardous to people, animals, agriculture, and property.[15] The biggest risks for Kilauea (2018) was the sulfur dioxide that can kill plants and cause respiratory failure in people and *laze* or lava haze that occurs when lava reaches the ocean and forms a dense white cloud of steam, toxic gas, and tiny shards of glass (Pele's hair).

EARTHQUAKES

Earthquakes occur when two earth surfaces move against each other. The Earth's crust is made up of plates that are moving in different directions and at speeds of fractions of inches up to five inches in a year. As the plates move against each other, tension develops along *faults*. A fault is a zone between two plates. The tension is periodically released suddenly when there is a slip along the fault line. The size of the surface area that moves when a fault line ruptures determines the amount of energy released.

When an earthquake occurs, the vibrations or seismic waves that emanate outward from the rupture can be measured on a seismograph. These machines can detect strong earthquakes from sources anywhere in the world. Earthquakes with magnitude of about 2.0 or less are usually called *microearthquakes*; they are not commonly felt by people and are generally recorded only on local seismographs. Magnitudes of about 4.5 or greater—there are several thousand such shocks annually—are strong enough to be recorded by sensitive seismographs all over the world. *Great earthquakes*, such as the 1964 Good Friday earthquake in Alaska, have magnitudes of 8.0 or higher. On the average, one earthquake of such size occurs somewhere in the world each year.[16]

---

[14] https://www.earthmagazine.org/article/benchmarks-may-8-1902-deadly-eruption-mount-pelee.

[15] https://volcanoes.usgs.gov/vhp/gas.html.

[16] https://earthquake.usgs.gov/learn/topics/measure.php.

MMS	Earthquakes/yr
8.5-8.9	0.3
8.0-8.4	1.1
7.5-7.9	3.1
7.0-7.4	15
6.5-6.9	56
6.0-6.4	210

magnitude 5 earthquake

magnitude 6 earthquake

magnitude 7 earthquake

FIG. 9.12   Frequency of major earthquakes worldwide over recent 47-year period of time, USGS. *(left) Courtesy of New Mexico Bureau of Geology and Mineral Resources. (right) Courtesy of USGS.*

Magnitude of an earthquake is determined using the Richter scale and expressed in whole numbers and decimal fractions. For example, a magnitude 5.3 might be computed for a moderate earthquake, and a strong earthquake might be rated as magnitude 6.3. Because of the logarithmic basis of the scale, each whole number increase in magnitude represents a 10-fold increase in measured amplitude. As an estimate of energy, each whole number step in the magnitude scale corresponds to the release of about 31 times more energy than the amount associated with the preceding whole number value.[17]

The Richter scale is not commonly used anymore, except for small earthquakes recorded locally. For all other earthquakes, the *moment magnitude scale* (MMS) is a more accurate measure of the earthquake size. The MMS has been in use since 2002 and is the scale used by the USGS to calculate and report magnitudes for all large earthquakes. Fig. 9.12 provides an approximate number of large earthquakes that occur each year worldwide.

Earthquakes are difficult to predict and impossible to prevent. The US Geological Survey (USGS) maps out high-risk zones based on seismic and geologic data that take into account when and where earthquakes occurred in the past, among other measures that help them predict future events (Fig. 9.13).[18]

California is known as an earthquake hot zone with more than 300 fault lines crossing the state. Particularly at risk is the southern part of the San Andreas Fault, which runs close to Los Angeles and has remained quiet since a magnitude 7.9 earthquake in 1857, meaning a lot of pressure has built up along that part of the fault line. According to USGS, Southern California, Los Angeles, and San Francisco Bay Area are most at risk in the coming years.

In recent years, the Cascadia subduction zone located in the Pacific Northwest has been a popular scenario for exercises as it is way past due for a major movement. This fault runs from Vancouver Island to northern California, and pressure has been building since the 1700s. An earthquake on this fault could produce a 9.0-magnitude shake.[19]

Alaska averages about 1000 earthquakes per month making it one of the most seismically active regions in the world. The state had the second-largest earthquake ever recorded in 1964 and has 11% of the world's recorded earthquakes. According to the state government, 7 of the 10 largest earthquakes in the United States were in Alaska, and one earthquake with a magnitude of eight or higher has occurred every 13 years since 1900.[20] According to the Oklahoma Geological Survey, Oklahoma experiences upwards of 600 3.0 earthquakes annually—much of which has been attributed to wastewater disposal (fracking).

---

[17] https://earthquake.usgs.gov/learn/topics/measure.php.

[18] https://earthquake.usgs.gov/hazards/hazmaps/.

[19] http://time.com/4949931/will-a-catastrophic-earthquake-strike-the-u-s/.

[20] http://seismic.alaska.gov/earthquake_risk.html.

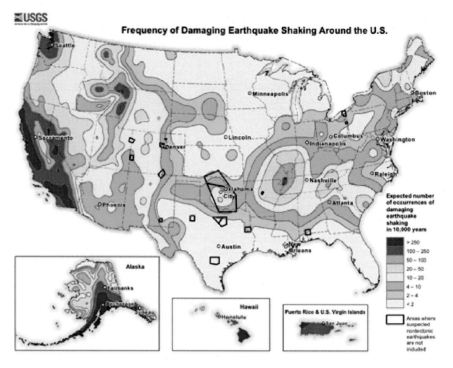

FIG. 9.13 USGS map of potential hot zones for earthquakes in the United States. *Courtesy of USGS.*

TSUNAMI

Tsunamis are typically caused by large, undersea earthquakes at tectonic plate boundaries. When the ocean floor at a plate boundary rises or falls suddenly, it displaces the water above it and launches the rolling waves that will become a tsunami. Most tsunamis—about 80%—happen within the Pacific Ocean's "Ring of Fire," a geologically active area where tectonic shifts make volcanoes and earthquakes common.[21] Tsunamis can also be caused by underwater landslides or volcanic eruptions. Tsunamis can travel up to 500 mi an hour and traverse the entire Pacific Ocean in less than a day without losing much energy!

In deep water, tsunami waves may appear only a foot or so high, but as they approach shoreline and enter shallower water, they slow down and begin to grow in energy and height. The tops of the waves move faster than their bottoms do, which causes them to rise to unbelievable heights. On March 11, 2011, a magnitude nine earthquake hit Northeastern Japan (Fig. 9.14), unleashing a savage tsunami felt around the world. The tsunami reached a height of 128 ft and traveled 6 mi inland. However, the unexpected disaster was neither the largest nor the deadliest earthquake and tsunami to strike this century. That record goes to the 2004 Banda Aceh earthquake and tsunami in Sumatra, a magnitude 9.1, which killed more than 230,000 people.[22]

A tsunami's trough, the low point beneath the wave's crest, often reaches shore first. When it does, it produces a vacuum effect that sucks coastal water seaward and exposes harbor and sea floors. This retreating of seawater is an important warning sign of a tsunami, because the wave's crest and its enormous volume of water typically hit shore 5 min or so later.

HAZARDOUS MATERIALS

Hazardous material (hazmat) spills occur on a regular basis in the United States. From motor vehicle accident, train derailments, and toxic ocean oil spills, hazmat accidents may be the most common type of disaster the country experiences. Recently (May 24, 2018), Seattle experienced an overturned 53′ trailer full of chicken feathers on I-5 at the start of the peak commute time. The DOT crew tasked with the cleanup of 40,000 lb of feathers resorted to PPE and breathing

[21] https://www.nationalgeographic.com/environment/natural-disasters/tsunamis/.

[22] https://www.livescience.com/39110-japan-2011-earthquake-tsunami-facts.html.

protection when the odor became nearly overwhelming. For all of the inquiring minds, the chicken feathers have a number of commercial uses including high-protein farm animal feed, enhancing plastic durability, and weather- and soundproof insulation. Accidents like this happen on the nation's highways multiple times every day. According to the Houston Chronicle, the Houston area saw more than 1000 heavy truck accidents in 2017—about 10% of those involved hazmat incidents.[23]

Hazmat events are categorized as I–III with a level III hazmat event being considered beyond the capabilities of the local hazmat material team and likely requiring large-scale evacuations. Examples include large releases from tank cars, train derailments, gas tank or chemical tank explosions, and weapons of mass destruction (WMD).

Two oil spills in US waterways have had significant impact on wildlife. On March 24, 1989, the Exxon Valdez hit a reef, and over 11 million gallons of oil was released contaminating 1300 mi of shoreline. The environmental consequences were devastating; estimates for animal impact included 250,000 seabirds, 2800 sea otters, 300 harbor seals, 250 bald eagles, and up to 22 killer whales killed, and billions of salmon and herring eggs were lost.[24]

On April 20, 2010, the largest marine oil spill in US history occurred in the Gulf of Mexico off the coast of Louisiana (Fig. 9.15). Natural gas traveled up the rig's riser to the platform where it ignited, killing 11 and injuring 17. The rig capsized and sank on the morning of April 22, and according to the US district court's findings of fact, approximately 134 million gallons of oil was released.[25]

In total, the oil spill likely harmed or killed approximately 82,000 birds of 102 species; approximately 6165 sea turtles; and up to 25,900 marine mammals, including bottlenose dolphins, spinner dolphins, melon-headed whales, and sperm whales. The spill also harmed an unknown number of fish—including bluefin tuna and substantial habitat for the nation's smallest seahorse—and an unknown but likely catastrophic number of crabs, oysters, corals, and other sea life.[26] The spill also oiled more than a thousand miles of shoreline, including beaches and marshes, which took a substantial toll on the animals and plants found at the shoreline, including seagrass, beach mice, and shorebirds.

TORNADOES

The United States gets its fair share of tornadoes each year—about 1200—which is more than any other country.[27] A good chunk of those tornadoes occur in what is referred to as Tornado Alley—an that is the area east of the Rocky

[23] https://www.simmonsandfletcher.com/blog/hazmat-accidents-increasing-texas-roadways/.

[24] http://www.findingdulcinea.com/news/on-this-day/March-April-08/On-this-Day–Exxon-Valdez-Captain-Acquitted-After-Oil-Spill-.html.

[25] http://www.gulfspillrestoration.noaa.gov/sites/default/files/wp-content/uploads/Chapter-2_Incident-Overview_508.pdf.

[26] http://www.biologicaldiversity.org/programs/public_lands/energy/dirty_energy_development/oil_and_gas/gulf_oil_spill/a_deadly_toll.html.

FIG. 9.15 Satellite image from Saturday June 26, 2010 showing oil leaking from the damaged Deepwater Horizon well in the Gulf of Mexico. *NASA image courtesy of the MODIS Rapid Response Team.*

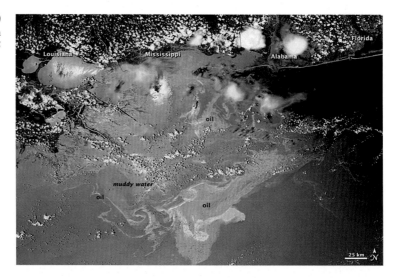

Mountains running from Texas to Canada. The Great Plains, the Midwest, and the Mississippi Valley are especially prone to tornadoes in the spring and early summer months. The southern region of the United States is another area that is hard hit including the northern parts of Alabama and Mississippi.

Peak tornado *season* is late spring through early summer (March to June) when cooler air coming over the Rocky Mountains meets warmer, moist air coming up from the Gulf. During the months of May and June, tornado activity is at its peak in the southern Great Plains (southern Kansas, Oklahoma, and Texas). Fig. 9.15 shows that tornadoes have occurred in the United States in every month of the year. Tornadoes in the winter time tend to occur in southern states as cold air pushed southward and meets warmer air at the Gulf (Figs. 9.16 and 9.17).

We also see tornadoes during the months of hurricane season in the Gulf Coast and Southeastern states. Hurricane Harvey (2017) produced at least 37 tornadoes, but the largest known outbreak of tropical cyclone tornadoes belongs to Hurricane Ivan (2004) that caused 117 tornadoes.[28] Interestingly, tornadoes are most likely to form in the right-front quadrant of a hurricane[29]—a good reason for responders to be familiar with the entire reach of a hurricane as it approaches landfall.

FIG. 9.16 Tornado frequency by month in the United States. *Image Credit: NOAA/National Centers for Environmental Information (NCEI).*

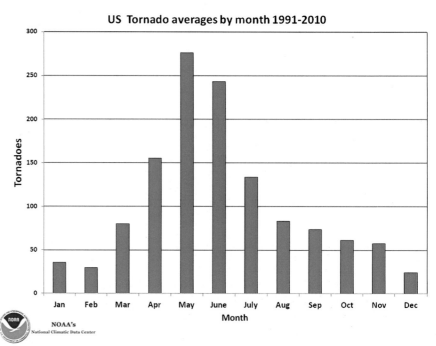

[28] https://weather.com/storms/hurricane/news/hurricane-harvey-tornado-reports.

[29] https://www.livescience.com/37235-how-hurricanes-spawn-tornadoes.html.

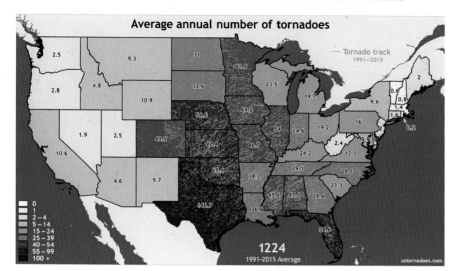

FIG. 9.17 Number of tornadoes on average by state. *Courtesy of* USTornadoes.com.

ANATOMY OF A TORNADO

Tornadoes form from an extreme struggle of hot and cold air. There's warm, moist air below and cold, dry air above, with a thin lid of stable air between. Sometimes, the warm air rushes through the lid of stable air and mixes with the cold air. An updraft and a downdraft begin and a thunderstorm forms. Air rotating on a horizontal axis gets pushed diagonally from the updraft, resulting in a tornado (Fig. 9.18).[30]

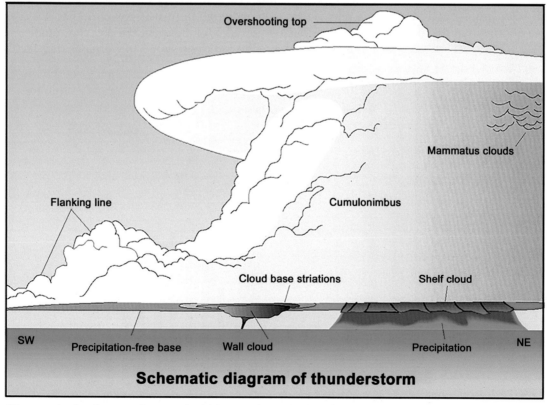

FIG. 9.18 Diagram of a thunderstorm. *Courtesy of NOAA National Severe Storms Laboratory.*

[30] http://tornadoscience.weebly.com/anatomy-of-tornados.html.

FIG. 9.19 The Enhanced Fujita scale. *Info-graphic courtesy of The National Weather Service/NOAA.*

Tornadoes are ranked by the damage that they do using the Enhanced Fujita scale (Fig. 9.19) that is a six-tiered assessment system that rates the tornado on the most intense damage within its path. Ninety-six percent of the states have seen strong/significant tornadoes (EF 2+) or greater. Only Alaska and Nevada have NOT reported tornadoes of this strength. Sixty-four percent of the states have been impacted by a violent tornado (EF 4–5). Forty percent of the states have been hit by an EF 5.

TIMING

Most tornadoes occur in the afternoon and evening hours. That is when most thunderstorms gain most of their energy from solar heating and heat that is released by the condensation of water vapor. That doesn't mean that we won't see a tornado in the morning—tornadoes have occurred at all hours of the day and, unfortunately, even in late evening when folks are sleeping and likely ill-prepared to rapidly respond to warnings (Fig. 9.20).

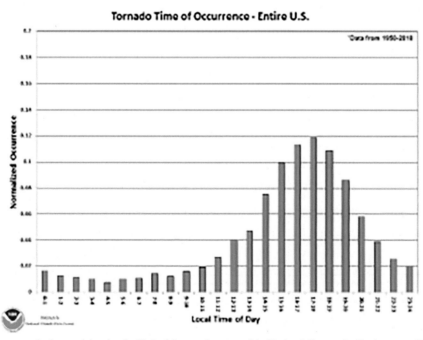

FIG. 9.20 Tornado occurrence by hour of day for the United States. *Courtesy of the National Centers for Environmental Information (NCEI)/NOAA.*

Billion-dollar events to affect the U.S. from 1980 to 2017 (CPI-Adjusted)

DISASTER TYPE	NUMBER OF EVENTS	PERCENT FREQUENCY	CPI-ADJUSTED LOSSES (BILLIONS OF DOLLARS)	PERCENT OF TOTAL LOSSES	AVERAGE EVENT COST (BILLIONS OF DOLLARS)	DEATHS
▪ Drought	25	11.4%	$236.6	15.4%	$9.5	2,993[†]
▪ Flooding	28	12.8%	$119.9	7.8%	$4.3	540
▪ Freeze	8	3.7%	$27.6	1.8%	$3.5	162
▪ Severe Storm	91	41.6%	$206.1	13.4%	$2.3	1,578
▪ Tropical Cyclone	38	17.4%	$850.5	55.3%	$22.4	3,461
▪ Wildfire	15	6.8%	$53.6	3.5%	$3.6	238
▪ Winter Storm	14	6.4%	$43.1	2.8%	$3.1	1,013
▪ All Disasters	219	100.0%	$1,537.4	100.0%	$7.0	9,985

FIG. 9.21 US Billion-dollar weather disasters: 1980–2016.

Tornadoes and hurricanes are the two mostly costly weather events in the United States. The Joplin Tornado in 2011 was the costliest tornado at close to $3 billion followed closed by the Tuscaloosa Tornado that occurred 1 month earlier. The two tornadoes in Moore, OK (2013, 1999), are certainly in the top 10 list with somewhere of $2.1 and $1.5 billion in losses, respectively (Fig. 9.21).[31]

NUCLEAR ACCIDENTS

Nuclear power plants operate in most states in the country and produce about 20% of the nation's power. Nearly 3 million Americans live within 10 mi of an operating nuclear power plant.[32] Nuclear power plants use the heat generated from nuclear fission in a contained environment to convert water to steam, which powers generators to produce electricity. An accident to a nuclear power plant could result in dangerous levels of radiation that could affect the health and safety of humans and animals living near the facility. The primary concern is a loss of cooling that could result in melting of the nuclear reactor core and a subsequent release of radioactive material.

There have been three major accidents to nuclear power plants over the last 40 years. The Three Mile Island reactor (Fig. 9.22) located near Middletown, PA, partially *melted down* on March 28, 1979 resulting in the most serious accident in a US commercial plant. Fortunately, only small amounts of radioactive material were released with no detectable health effects for the workers or the public. Follow-up studies by a number of "well-respected" agencies found no lingering effects on plant or animal life, but the accident had a major impact on regulatory oversight and enhancing reactor safety.

Seven years later, on April 26, 1986, a sudden surge of power during a reactor system test destroyed unit 4 of the nuclear power station at Chernobyl, Ukraine. The accident and the fire that followed released massive amounts of radioactive material into the environment. Thirty-one people died immediately, and hundreds were sickened. Over 335,000 people were evacuated.[33]

Overall, there was an increase in mortality and a decrease in reproduction for animals located in the exclusion zone (Fig. 9.23). During the first few years after the accident, plants and animals showed many genetic effects of radiation. Still today, there are reports of anomalies in plants and animals both in the exclusion zone and beyond. As the radioactivity levels have decreased over the last 30 years, biological populations have been recovering from the acute radiation effects. Following the initial reductions in numbers, some of the populations have recovered and grown, and since human activity has stopped, the exclusion zone has become a unique sanctuary for biodiversity.[34]

[31] http://www.spc.noaa.gov/faq/tornado/damage$.htm.

[32] https://www.ready.gov/nuclear-power-plants.

[33] https://www.nrc.gov/reading-rm/doc-collections/fact-sheets/chernobyl-bg.html.

[34] https://www.greenfacts.org/en/chernobyl/l-2/3-chernobyl-environment.htm.

FIG. 9.22 Three Mile Island personnel cleaning up the contaminated auxiliary building. *Image taken from Report of the President's Commission on the Accident at Three Mile Island.*

FIG. 9.23 Second generation of Chernobyl dogs. *Photo courtesy of Kyle Held.*

On March 11, 2011, an 8.9 major earthquake in Northeastern Japan took out the primary power supply to the Fukushima Daiichi nuclear power plant. Less than an hour later, a massive tsunami disabled the backup generators. Rising residual heat within each reactor's core caused the fuel rods in reactors 1, 2, and 3 to overheat and partially melt down, leading to a release of radiation.[35] Although the quantity of radioactive noble gases released from Fukushima exceeded the amount released from Chernobyl, the size of land area severely contaminated by ^{137}cesium (^{137}Cs) was 10 times smaller around Fukushima compared with around Chernobyl.[36]

Environmental impact has been studied for terrestrial and marine environments. Studies of plants and animals in the affected area have recorded a range of physiological, developmental, morphological, and behavioral consequences

[35] https://www.britannica.com/event/Fukushima-accident.

[36] https://academic.oup.com/jrr/article/56/suppl_1/i56/2580293.

FIG. 9.24   Screening test for radioactive contamination carried out by Hirosaki University staff. *Journal of Clinical Biochemistry and Nutrition, 50(1), 2–8. https://doi.org/10.3164.*

of exposure to radioactivity. Various effects have been observed in the exposed populations of Fukushima monkeys, butterflies, birds, trees, and aphids. Sampling in the marine environment that assessed radiation exposure to marine birds indicated doses below the thresholds for detection radiation effects in these populations. When compared with the Chernobyl accident, the impact of the Fukushima accident seems to be far less severe.[37]

Both the Fukushima and Chernobyl accidents were rated at the maximum level (level 7) on the International Atomic Energy Agency (IAEA) nuclear event scale,[38] indicating an accident with large release of radioactivity accompanied by "widespread health and environmental effects" (Fig. 9.24). However, there are a number of significant differences between Chernobyl and Fukushima. The amount of the release at Fukushima was about 10% of that at Chernobyl; the presence of the Fukushima containment structures, the radionuclides released (mostly iodine and cesium isotopes vs the entire core inventory at Chernobyl), the physical form of the releases (mostly aqueous vs volatile), the favorable currents and winds at the site, and the timing of the release with respect to population evacuation all resulted in vastly smaller overall consequences for Fukushima.[39]

Approximately 60,000 people were evacuated following the Fukushima accident, and about one-half have returned. Many of those that were ordered to evacuate were not able to take their animals. The livestock could not be sold, and a number of the farmers did not want to kill them. Many domestic pets were left behind as well to fend for themselves. The International Fund for Animal Welfare convened a group of subject matter experts in Tokyo on May 2 and 3 and from that 2-day summit provided the government of Japan recommendations for the monitoring, safe rescue, and decontamination of pets, livestock, and wildlife.[40] Unfortunately, many of the animals impacted were left to die or were at the mercy of the few residents who come in daily to care for them. A number of rogue individuals and rescue groups (primarily from outside of Japan) did sneak into the evacuation zone to rescue animals. Their goal was to provide temporary sheltering until transport could be arranged to the United States—a plan that was quickly thwarted by a number of US agencies.

The US Nuclear Regulatory Commission (NRC) licenses and regulates commercial nuclear power plants; research, test, and training reactors; nuclear fuel cycle facilities; and the use of radioactive materials in medical, academic, and industrial settings. The NRC also regulates the transport, storage, and disposal of radioactive materials and waste.[41] The NRC has developed four emergency classification levels for alerting the public on potential or actual consequences from a nuclear event[42]:

---

[37] https://www.elsevier.com/connect/5-years-after-fukushima-insights-from-current-research.

[38] https://www.iaea.org/topics/emergency-preparedness-and-response-epr/international-nuclear-radiological-event-scale-ines.

[39] http://web.mit.edu/nse/pdf/news/2011/Fukushima_Lessons_Learned_MIT-NSP-025.pdf.

[40] https://www.ifaw.org/united-states/resource-centre/nuclear-accidents-and-impact-animals.

[41] https://www.nrc.gov/reading-rm/doc-collections/nuregs/brochures/br0099/.

[42] https://www.nrc.gov/reading-rm/doc-collections/fact-sheets/incident-response.html.

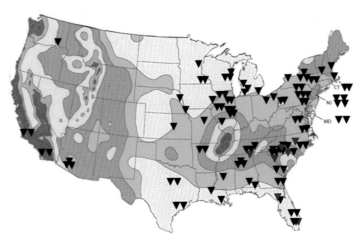

FIG. 9.25 Seismic activity and location of nuclear power plants. Highest risk denoted purple followed by red. Moderate risk in orange followed by yellow and green. Low risk in blue and gray. *Image courtesy of The Center for Public Integrity (https://www.publicintegrity.org/2011/03/18/3700/). The Center for Public Integrity is a nonprofit, nonpartisan newsroom based in Washington D.C.*

Notification of unusual event—Events are in process or have occurred that indicate a potential decline in the level of safety of the plant. No release of radioactive material requiring off-site response or monitoring is expected at that time.

Alert—Events are in process or have occurred that involve an actual or potentially substantial decline in the level of plant safety. However, any release of radioactive material is expected to be only a fraction of the Environmental Protection Agency (EPA) protective action guidelines.

Site area emergency—Events are happening or have occurred that involve an actual or potential major failure of the plant's ability to protect the public. Any releases of radioactive material are not expected to exceed the EPA guidelines except near the site boundary.

General emergency—Events are in progress that involve actual or imminent severe damage or melting of radioactive fuel in the reactor core. There is a potential for radioactive releases exceeding EPA guidelines beyond the immediate site area.

Fig. 9.25 combines a USGS map of earthquake activity zones with nuclear power plant sites, and there are a number of plants that could be located in areas prone to seismic activity.[43] Similar threats of hurricanes, tornadoes, and fires would reveal similar concerns for plant and community safety.

NUCLEAR DETONATION

A nuclear detonation would result in a catastrophic disaster in a major urban area, with enormous loss of life; high numbers of injuries; destruction of infrastructure; and radiological contamination of significant areas, including outlying agricultural areas, and would require tremendous response and recovery resources. The FEMA chemical, biological, radiological, nuclear, and explosive (CBRNE) branch has engaged in a planning project to identify workable response and recovery strategies and highlight and work toward remediation of capability gaps. An animal workgroup composed of federal, state, local, and nongovernmental stakeholders was formed to assist the CBRNE branch in their planning efforts. The purpose of this group was to provide an analysis of animal-mission coordination and resource management and offer consensus recommendation on actions to meet the significant resource challenges posed by such an incident.[44]

The immediate and overwhelming nature of this type of incident will produce initial and critical resource shortfalls in almost every aspect of response in the midst of extreme chaos. Some animal response missions will need to wait until significant progress has occurred in human life-safety response. The movement of animal resources (personnel,

---

[43] https://www.publicintegrity.org/2011/03/18/3700/regulators-aware-years-understated-seismic-risks-nuclear-plants.

[44] FEMA CBRNE Animal Work Group. 2017.

supplies, and equipment) for this type of response will likely take some time to arrive given the prioritization of human life-saving efforts. Immediate needs for animal response might include the following:

(1) Multiagency coordination of available animal response resources
(2) Coordination and support of public messaging pertaining to animal issues
(3) Support of mass care mission pertaining to people with pets, service animals, and other animals:
    (a) Pet and service animal evacuation support
    (b) Pet and service animal sheltering
    (c) Veterinary medical care for pets and service animals
    (d) Pet and service animal decontamination
    (e) Shelter in place support for families with pets and service animals
(4) Rapid initiation of agricultural movement controls to ensure that contaminated agricultural items, including livestock and poultry, do not enter the food system.[45]

[45] FEMA CBRNE Animal Work Group. 2017.

10

Animal Behavior

The majority of animals rescued in the United States following a disaster will be cats and dogs. But in rural areas, the ASAR team may be asked to rescue livestock and even fish and wildlife. The rescue techniques vary greatly between species and situations. This section will discuss some basic considerations for the most common requested species.

DOGS

Every breed of dog will react differently and unpredictably when stressed. Some will become more aggressive, while others may shut down and cower. Without a doubt, there will still be some commonalities among breeds—the typically friendly golden retriever may still react in a gentle way, and the nipping, ankle-biting Chihuahua may be still looking for a way to show his distaste for being handled by a stranger. The fact is that you never know how a dog will react and you should not base your rescue strategy solely on the breed or size of the dog. Nor should you base your strategy on an owner's description of a pet's behavior. Pets will react differently when stressed, so the responder needs to be cautious even when the owner assures them that their pet is a true sweetheart as seen in this photo (Fig. 10.1).

Plan for every shape, size, and temperament before responding to a call. Make sure that you have carriers that will accommodate the dog and be of adequate size and containment quality. If the dog is indeed gentle and accommodating, then a lead may be all that is needed. If however the dog is large and aggressive, a catchpole may be the preferred method. Regardless of the rescue equipment, the rescuer's safety is number one. If the dog appears to be overly aggressive, call for backup, and if there are doubts whether the animal can be rescued safely, *back out and rethink the situation*. Dogs may bite because they feel threatened, or they may bite because they are in pain and sick or in order to protect their area, family, food, or even a toy.

In addition to an understanding of various breeds and an appreciation of size (large dogs are typically stronger and closer to the responder's upper body—small dogs are typically held, which places them closer to the face) and gender (70%–76% of dog bites come from intact males),[1] there are other signs that the dog may exhibit to give the rescuer an indication of how to approach and restrain the animal. The ASPCA[2] has developed a list of behaviors that might indicate aggression:

- Becoming very still and rigid
- Guttural bark that sounds threatening
- Lunging forward or charging at the person with no contact
- Mouthing, as though to move or control the person, without applying significant pressure
- "Muzzle punch" (the dog literally punches the person with her nose)
- Growl
- Showing teeth
- Snarl (a combination of growling and showing teeth)
- Snap
- Quick nip that leaves no mark

[1] CDC.

[2] https://www.aspca.org/pet-care/dog-care/common-dog-behavior-issues/aggression.

- Quick bite that tears the skin
- Bite with enough pressure to cause a bruise
- Bite that causes puncture wounds
- Repeated bites in rapid succession
- Bite and shake

You can tell a lot about a dog's emotional state by understanding their body language. Body language in dogs, much like humans, may help in determining whether it is safe to approach or whether the dog is highly stressed. A dog that has their tail down and relaxed, ears up, mouth opened slightly with the tongue exposed, and the head high is typical of a dog that is relaxed, unthreatened, and likely approachable (Fig. 10.2). Following photos from Pexels.[3]

A dog that is alert and aware of something of interest or a potential threat may have his tail more horizontal, ears will be forward, eyes are wide, mouth is closed, and he may be leaning forward (Fig. 10.3).

When a dog's tail is raised and bristled, hackles are raised, ears are forward, nose is wrinkled, lips are curled, and body position is forward, then you have the *potential* for an aggressive, dominant, and likely a confident dog that might act aggressively if challenged (Fig. 10.4). Dogs will charge another animal or human intruder—oftentimes without exhibiting aggressive signs or without being provoked. Consequently, the responder must always approach slowly and carefully when the dog is not known or when it is highly stressed and/or injured. Photo from iheartdogs.com.[4]

Dogs can be fearful and aggressive. In that situation, tail is tucked, body is lowered, ears are back, pupils dilated, and nose may be wrinkled. This is a dog that is frightened but may attack if pushed (Fig. 10.5). Photo from Daxton's Friends.com.[5]

FIG. 10.2 Relaxed and approachable dog. *Courtesy of Pexels.*

[3] https://www.pexels.com/search/dog/.

[4] https://iheartdogs.com/researchers-identify-possible-risk-factors-for-anxiety-aggression-in-dogs/.

[5] http://www.daxtonsfriends.com/identifying-dangerous-behavior/types-of-dog-aggression/.

FIG. 10.3 Alert dog. *Courtesy of Pexels.*

FIG. 10.4 Potentially aggressive dog. *Courtesy of* iheartdogs.com.

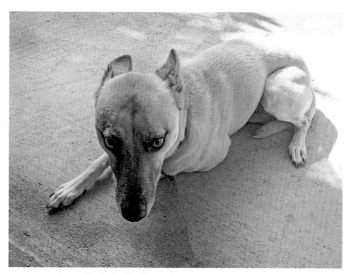

FIG. 10.5 Frightened dog. *Courtesy of Ellen Levy Finch. Wikimedia Commons.*

FIG. 10.6 Submissive dog. *Courtesy of The American Society for the Prevention of Cruelty to Animals. All rights reserved.*

Signs that indicate a dog has surrendered and is submissive may include lying on their back, ears flat and back, head turned, and sprinkling some urine (Fig. 10.6). This is a dog that is hoping to avoid any contact or confrontation and if handled properly will not become aggressive. But careful handling is imperative as a very frightened animal may strike out—thus the phrase fear-biter. Photo from Dog Training Excellence.[6]

As mentioned previously, "when in doubt, back out," but dogs can sense when their "rescuer" is unsure or lacks confidence. In general, avoid eye contact, present a smaller frame, and avoid putting yourself in a dangerous position. If you are going to offer a hand for sniffing, make sure it's a fist rather than exposed fingers. Avoid grabbing a dog from an owner/responder if at all possible—the move from owner to stranger can set off an aggressive response. If a handoff is required, receive the dog with the front end facing the initial holder—butt bites are much less painful.

Avoid petting an unknown dog as it may become overstimulated. Consider a basket muzzle when appropriate. Basket muzzles allow the dog to pant, which helps with cooling. Field muzzling with rope or other material should only be used for emergencies and only for short periods of time.

There are a number of essential rescue devices available for handling dogs. The most common device is a simple slip lead. That may be the only piece of equipment needed to bring an animal under control and to move safely. Leads need to be sturdy, long enough to provide some distance if needed, and strong enough to prevent escape. Nylon slip leads may be ideal in a sheltering environment but for field work, a cable lead or a 3/8″ minimum, polypropylene rope with some form of stop is recommended.[7] The rescuer should also be trained in using nets and catchpoles (rabies stick). The 5′ catchpole shown here (Fig. 10.7) is ideal for capturing and restraining an aggressive animal.[8]

During the Haiti earthquake response in 2010, one of the US-based responders was working with the vaccination team when an owner came up with a puppy in her arms. She handed the puppy to the responder so that he could inspect the pup and determine if it was healthy. The responder instinctively accepted the pup. Moments later, the pup's mom came racing down the driveway and lunged at the responder and bit him in the groin area. It happened so quickly that there was no time to defend himself or hand the puppy back. The responder was treated and the owner was asked to not take the dog out of the community for 10 days so that the team could inspect the mom for signs of rabies. The responder was traumatized—an emotional wreck—and was sent home. At the end of 10 days, the team went back to see the mom, and the family had moved with no way of finding them. The responder went through the rabies postexposure vaccination process—an ordeal that is rather painful and expensive. As seen in Fig. 10.8, dog bites can be painful and debilitating.

[6] http://www.dog-training-excellence.com/barking-dog-problem.html.

[7] https://www.mendotapet.com/.

[8] http://www.animal-care.com/product/ketch-all-catch-pole/.

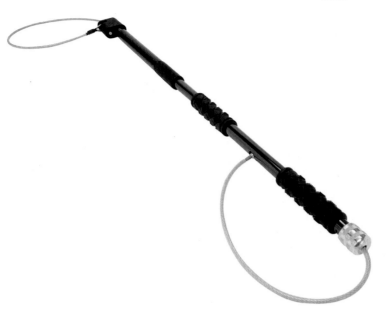

FIG. 10.7 Catch-pole.

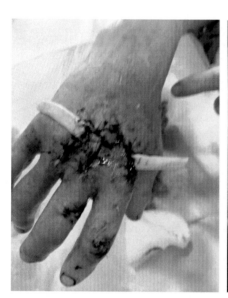
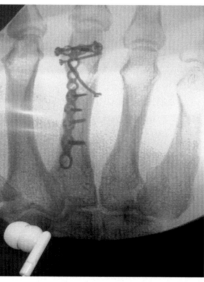

FIG. 10.8 Animal bite to the right hand of veterinarian. *Courtesy of MA Littlefield, DVM, MS, CVA.*

According to the Center for Disease Control and Prevention (CDC), 4.5 million people are bitten by a dog annually, and one in three will require medical attention.[9] Since it is oftentimes difficult to determine the breed of a dog, the CDC stopped keeping breed-specific dog bite statistics in 1988.[10] But that has not kept groups and concerned citizens from drafting breed-specific legislation (BSL) banning a particular breed in their community. The most targeted breed is the pit bull, and Fig. 10.9 shows that states have counties with BSL.[11]

Regardless of the type of dog that bites, the rescuer needs to complete an incident report to ensure proper documentation and make sure the animal is quarantined for a minimum of 10 days—even if there is proof of rabies vaccination. Wash the wound, keep clean, and seek medical attention immediately.

[9] https://www.cdc.gov/features/dog-bite-prevention/index.html.

[10] https://www.avma.org/News/JAVMANews/Pages/171115a.aspx.

[11] https://www.avma.org/News/JAVMANews/Pages/171115a.aspx.

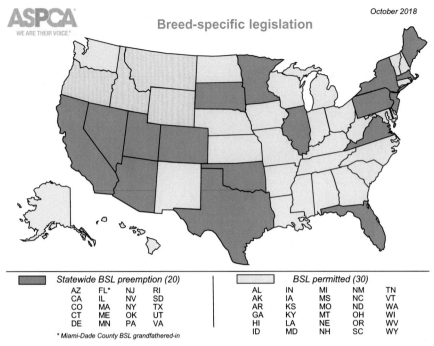

Breed-specific legislation

October 2018

Statewide BSL preemption (20)				BSL permitted (30)				
AZ	FL*	NJ	RI	AL	IN	MI	NM	TN
CA	IL	NV	SD	AK	IA	MS	NC	VT
CO	MA	NY	TX	AR	KS	MO	ND	WA
CT	ME	OK	UT	GA	KY	MT	OH	WI
DE	MN	PA	VA	HI	LA	NE	OR	WV
				ID	MD	NH	SC	WY

* Miami-Dade County BSL grandfathered-in

FIG. 10.9 States that have breed-specific legislation. *Courtesy of The American Society for the Prevention of Cruelty to Animals. All rights reserved.*

CATS

Cats can bite and scratch and are very good at both! Cat bites can be very painful and easily infected, and because they are so agile and quick, they can be difficult to handle when stressed. There are control devices to help constrain a cat, but much like the catchpole, it requires getting close enough to engage the device. Nets can be effective in having a greater reach but regardless of the tool, be prepared for a challenging experience if the cat is not friendly or socialized. They can climb furniture and gyrate their bodies and if you don't shut down their escape route, will be long gone by the time you finish calling "here, kitty"!

Many behaviorists and veterinarians do not favor scruffing a cat, but seasoned animal handlers will use this technique for aggressive cats. Scruffing entails one hand grabbing the cat behind the neck where the skin is loose, slightly extending the wrist to hyperextend the cat's spine, and placing the forearm along the cat's spine. The other hand can reach down and grab both legs and extend them (Fig. 10.10).[12] Obviously, this is an aggressive maneuver and not necessary for a friendly cat but still a recommended hold if there is any doubt. Male, intact cats (tom cats) may be more difficult to scruff as there is not the laxity in the skin on the back of the neck to get a good hold. Other restraint methods include wrapping an animal in a towel[13] and cat bags (Fig. 10.11).

Pole and drop nets, cat nabbers, and cat traps are valuable pieces of equipment for the cat rescuer. Do not make contact with an unknown cat until you have a carrier ready as you want to limit the amount of time you have to hold the cat.

Cat gloves are specifically designed and manufactured to prevent teeth and claws from reaching the skin. You can find bite gloves in a number of styles and arm lengths but with protection comes a loss of control as the gloves are thick and may make it more difficult to achieve a patent scruff. Seasoned responders may opt for a lighter handling glove that provides some protection but allows for greater dexterity and sensation.

There are a number of transport carriers that work well with cats with the most common being plastic and soft-sided crates. As with dogs, the carrier must be large enough to get the animal in and for them to be able to stand and make a complete turn. Unsocialized or stressed cats may benefit from a cozier, darker transport cage. If crates are not available, the responder can use a pillowcase or cat sack for short transports. If the cat is taken from the home, more than likely they have spent time on a bed (or two) and may find comfort in the friendly smells that come with the pillowcase they have spent time on.

[12] http://infovets.com/books/feline/B/B300.htm.

[13] https://www.veterinaryteambrief.com/article/techniques-towel-restraint-cats.

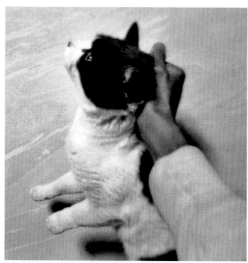

FIG. 10.10 Cat scruffing technique. *Courtesy of Wikimedia Commons/Romanee.*

FIG. 10.11 Immobilizing a cat with a tight towel wrap. *Courtesy of The American Society for the Prevention of Cruelty to Animals. All rights reserved.*

Cats can be very vocal from the more friendly meow to a blood-curdling yowl. The meow sound is typically a greeting of sorts or a way for them to gain your attention. Some cats will also trill or let out small chirps in attempt to steal your attention. Most cats will purr when they are content, but a cat may also purr if it is stressed or sick. But in general, these sounds are welcomed from a responder's perspective—it's the growling, hissing, or spitting that will place the rescuer on edge. This is a cat that does not want to be touched or bothered, and you may want to bring in the cat handlers. Cats that yowl may be in distress, or you may hear that sound from intact cats that are in search of a mate. And finally, cats that spend time on the window sill watching birds or chipmunks will oftentimes make a chattering or guttural trill sound—a sign that they would much prefer to be on the other side of the window chasing their prey.

The body language of a cat may provide some insight on their overall well-being. Unlike some dogs, cats will seldom charge another animal or human. They will strike out and defend themselves when cornered but would much prefer to hide when frightened than confront a rescuer. Cats will use their entire body to signal their intentions, but the rescuer needs to remember that cats are somewhat unpredictable and their body language may be conflicting or contradictory. The classic example is the cat that jumps up into your lap, lays down, and starts purring within moments of your petting and, then, for no apparent reason, rears back and bites or scratches you because at that instant, they had had enough.

Tails

A tail that is relaxed and up and fluffy is an approachable cheerful cat. If the tail is down, the cat may be scared or feel threatened. If the cat starts to thump its tail, it is telling you that it has had enough and is agitated. A cat slow

FIG. 10.12   Frightened cat. *Courtesy of Wikimedia Commons/Luis Miguel Bugallo Sánchez from Santiago de Compostela, Galicia].*

FIG. 10.13   Frightened cat. *Courtesy of Wikimedia Commons/ Peretz Partensky from San Francisco, USA.*

moving his tail back and forth may indicate that the cat is contemplating the situation, and a cat that has his tail straight up and fluffed out is an unhappy camper and is appearing to be as large and scary as he can be (Fig. 10.12). Photos from animalmozo[14] and Virily.[15]

Ears

Ears that are slightly forward may indicate a contented or even playful cat. When someone or something occurs within a cat's environment, the ears will often perk straight in attention. When the ears turn to the side or back as in the previous photo, this may indicate a cat that feels threatened, irritated, or overstimulated (remember cat in the lap story!). And if the ears are pressed flat against their head (see photo above), this is a sure sign that the cat feels threatened, and the rescuer needs to be prepared for a cat that might strike out.

[14] http://animalmozo.com/2016/02/18/people-shouldnt-scare-their-cats-with-cucumbers/.

[15] https://virily.com/entertainment/hillarious-scared-cats-compilation/.

Eyes

Cat (and human) pupils will change size with changes in light, but cat pupil size may also indicate levels of concern. Some cats will have their pupils dilated when they are surprised or scared. Constricted pupils may indicate that the cat is feeling aggressive. The rescuer should avoid a staredown with any animal as most species find it challenging, and cats are no exception. If the cat has a slow blinking action or has droopy, half-closed lids, it is likely relaxed and approachable (Fig. 10.13).

FISH

Fish are the most popular pet in the United States with nearly twice as many fish as dogs. Over 480 million goldfish are sold each year (Census of Aquaculture, NASS 2006),[16] and approximately ½ million koi are bred annually.[17] Prized koi can reach hundreds of thousands of dollars, and during the recent northern California fires (Sep., 2017), a breeding pair of koi were rescued and valued at $100,000 each! In warmer parts of the country, backyard ponds abound, and many are home for pet fish.

In 2017, about 139.3 million freshwater fish were owned by American households. Thus, freshwater fish made up the most popular pet category in the United States, based on the total number.[18] Whether fish are housed inside or outside, they will need electric power to keep filters and aerators running. Individuals that own large aquariums full of exotic, expensive fish will likely be calling animal control for assistance if long-term power outages are expected or in situations where access is being denied. If sheltering in place is not an option, the fish will need to be evacuated and that brings a whole host of challenges for the rescue team. Rescue may be as simple as strapping down a lid on a small aquarium and transporting to somewhere that has power to as extensive as the "no koi left behind" operation following the Tubbs Fire (2017) in Sonoma County, CA.

In Santa Rosa, where the homes were completely destroyed and the only power source was generators, the only viable solution was to relocate the koi. Sonoma County Animal Services freed up space in a warehouse adjacent to their shelter. The ASPCA contacted a koi club in the bay area who in turn contacted a number of other clubs throughout northern California and were able to locate, transport, and set up show ponds in the warehouse. SCAS put in a request to the EOC for a water tender to fill the ponds. Koi do not do well in chlorinated water, so the water was dechlorinated and aerators were installed. A commercial fish company that specialized in moving live fish throughout the bay area agreed to bring their truck to Santa Rosa. On their truck, they had four 600 gal tanks. The local fairgrounds had a water pump that provided nonchlorinated water from a pond. The field teams met the transporter at the fairgrounds and over the next two days, moved koi from backyard ponds to the warehouse. Koi are very susceptible to disease, so the pond were segregated by property (Fig. 10.14).

To ensure that no koi was left behind, the rescue team used nets to capture the fish. The first few sweeps typically yielded a good number, and after that, the remaining fish headed for the bottom so the teams installed pumps to drain the ponds and the process was repeated until every fish was captured (Fig. 10.15). No fish were lost during the entire process, which says a lot about the rescue team's efforts but says even more about the amazing collaboration between the various groups that chipped in to make it happen. There was some concern initially that the community would not be entirely receptive to the efforts to save the koi, and had those efforts been started earlier, that might have been the case. But because the teams waited until the human and four-legged animal needs had been addressed, the timing was right for this operation, and it was also the right time for a community so devastated by the fires to hear a feel good story.

Shortly after the koi experience in Santa Rosa, the ASPCA was requested to assist County of Santa Barbara Animal Services following the mudslides (Jan. 2018). A large number of homes in the Montecito area had backyard fish ponds, but unlike Santa Rosa, the majority of the homes that had ponds were not affected by the slides, and the power was slated to come back on over the next several weeks. So in this case, Santa Barbara County Animal Services opted to shelter in place and agreed to provide transportation into the restricted areas for a contractor to care for the fish until the owners returned.

[16] http://articles.extension.org/pages/58765/goldfish.

[17] https://pethelpful.com/fish-aquariums/Koi-Carp-An-Ancient-Long-Lived-Ornamental-Fish.

[18] https://www.statista.com/statistics/198104/freshwater-fish-in-the-united-states-since-2000/.

FIG. 10.14   Koi ponds. *Sonoma Fires, 2017. Courtesy of The American Society for the Prevention of Cruelty to Animals. All rights reserved.*

FIG. 10.15   No koi left behind. *Sonoma Fires, 2017. Courtesy of The American Society for the Prevention of Cruelty to Animals. All rights reserved.*

CAPTIVE WILDLIFE

There are more than 2400 USDA licensed animal exhibitors in the United States,[19] ranging from very large facilities to private individuals with few animals. Regardless of the size of the facility, given the inherent risk associated with captive wildlife, it is imperative that all the venues prepare for disasters and have contingency plans in place. One

[19] https://www.whyanimalsdothething.com/who-are-the-usda-class-c-exhibitors/.

FIG. 10.16 Miami Zoo flamingos sheltering in place in men's room. *Courtesy of Ron Magill.*

hundred seventy-five million people visit zoos or aquariums annually; therefore, emergency response planning must take the welfare of visitors, staff, first responders, collection animals, agricultural animals, and even local wildlife into consideration.[20]

When captive wildlife facilities are damaged by disasters, the community and the animal will be at risk. Muskingum County Animal Farm was a private zoo in Zanesville, OH, owned by Terry Thomson, a Vietnam veteran. On October 19, 2011, before committing suicide, he released 56 animals including big cats, bears, and wolves. Forty-eight of those 56 animals were eventually killed by local police. The remaining animals were tranquilized and sent to the Columbus Zoo. This was not a natural disaster but certainly demonstrates what could happen if a zoo was struck by a disaster and the inhabitants entered neighboring communities.

In late May 2011, two zoos in central North Dakota were hit hard by flooding requiring the evacuation of the animals. In Bismarck, the Missouri River threatened to submerge the Dakota Zoo and its 500 + animals under as much as 7 ft of water, and in Minot, the Roosevelt Park Zoo was a potential target of the rising Souris River, which runs through the city. The Roosevelt Park Zoo got all its animals out in a 12 h period with all but the bears being loaded without chemical immobilization.

Moving captive wildlife is a huge undertaking, which is why many zoos have decided to shelter in place if a disaster is looming. These facilities recognize that they may lose an animal as a result of the disaster but the risks associated with moving wild animals is much greater than the risk of losing them to a disaster. Prior to landfall of Hurricane Georges on Sep. 25, 1998, Miami Metrozoo (now Zoo Miami) rounded up more than 50 Caribbean flamingos and kept them sheltered in place in a men's restroom—a tactic that they had also used before with Hurricane Andrew (Fig. 10.16).

Whether the plan is to evacuate or to stay put, captive wildlife facilities must have a plan and train that plan regularly to ensure the safety of the animals, staff, and patrons. The Association of Zoos and Aquariums represents more than 230 animal care facilities in the United States and abroad and requires all of its members to practice an annual disaster preparedness drill to keep their accreditation.

The two greatest risks for facilities holding wildlife in the state of CA is earthquakes and fires. Over the last several years, a number of smaller licensed facilities have had to evacuate their wild animals as a result of fires. Utilizing resources from the facility, local animal control, and state agriculture, the majority of the evacuations have gone reasonable well.

Fortunately, zoos are fairly fire-resistant with a good deal of concrete, and the larger facilities have excellent plans that are reviewed and practiced annually. In 1993 and 2003, wildfires threatened the San Diego Zoo. In 1993, the fire came right up to their fence line. Officials caught the park's condors and moved them away from the fire, but large animals did not have to be evacuated. In the 2003 fire, the park moved some birds and baby animals to safer quarters

[20] https://phys.org/news/2014-11-emergency-preparedness-zoos-aquariums.html.

but left the large animals in their expansive habitats where they could attempt to move away from smoke if necessary.[21]

The role of the animal responder in captive wildlife evacuations and emergencies is limited unless they have been specifically trained and have adequate safety measures in place. Recently, Code 3 Associates has provided training for zookeepers on how to utilize large animal rescue equipment for large wildlife extrication. And on a similar track, as a result of some of the lessons learned from Hurricane Harvey, the Texas Animal Health Commission has developed a collaborative working committee composed of zoos, aquariums, emergency management, academia, animal advocates, and agriculture to address some of the challenges facing zoos and captive wildlife facilities when disaster strikes.

[21] http://articles.latimes.com/2007/may/29/local/me-zoos29.

CHAPTER

11

Animal Search and Rescue

The NASAAEP Animal Search and Rescue (ASAR) Best Practice Working Group[1] defines ASAR as an operation mounted by emergency services, animal care and control, and trained volunteers, to locate, find, evacuate, and extricate animals believed to be in distress, lost, sick, stranded, trapped, or injured. ASAR operations may include the following:

1. Locating, evacuating, and extricating animal victims
2. Documentation of the found animals (where, when, status, disposition of animal, etc.)
3. Triaging and providing emergency/stabilizing medical treatment for animals located and found in the field
4. Coordinating animal rescue efforts
5. Confinement, capture, packaging, and transport of animals believed to be in distress to a safe and stable location

ASAR teams may be composed of the following:

- Animal rescue technicians who've been trained with the appropriate human SAR competencies
- Human SAR team members who have animal handling competencies
- Human/animal SAR integrated team

ASAR is a new member of the Search and Rescue community—gaining recognition and acceptance since Hurricane Katrina (2005). Even though various groups were performing animal search and rescue activities well before Katrina, the title did not gain traction until the multitude of debriefs, meetings, workshops, and trainings that followed the great storm. Interestingly, the animal welfare community was reluctant in adopting the title. They were the first to recognize that they did not have the training or the expertise that their human counterparts (US&R) had in the highly specialized, technical situations found in compromised structures or events requiring specialized training or equipment to affect a safe rescue. But what they did have were the requisite animal handling skills and a mission to rescue animals (Fig. 11.1). Consequently, in the months and years following Katrina, the concept of ASAR slowly but surely took hold.

That's not to say that there was immediate buy-in by all of the rescue community. Fig. 11.2 demonstrates the various groups/stakeholders that were either directly or indirectly related to ASAR.

There are a number of recognized SAR teams in addition to Animal Search and Rescue including wilderness, water, urban, and air search and rescue. In smaller communities, the local SAR team may be cross-trained to be able to handle most environments, and in larger communities, there may be specialized teams for each of the settings listed above. Wilderness SAR typically includes the use of search dogs and covers large geographic areas oftentimes of difficult terrain and features. Wilderness SAR teams are generally composed of folks that are comfortable being out on their own for several days and are well trained in navigation, wilderness survival, and emergency first aid.

Response teams may be composed of volunteers that have trained with emergency management and fire/rescue squads. In communities where there are a large number of rivers and/or lakes, there may be a dive recovery team, typically composed of local divers that have been trained in body recovery.

Urban SAR teams are highly trained teams that are well suited for entering conditions where structures or the environment is not stable. In addition to being able to organize and carry out a search procedure, these teams are capable of shoring up or supporting compromised spaces. In 1989, FEMA started the National US&R program.

---

[1] www.nasaaep.org.

FIG. 11.1   IFAW ASAR team, 2008. *Courtesy of The International Fund for Animal Welfare. All rights reserved.*

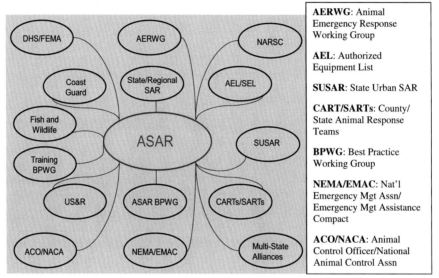

FIG. 11.2   ASAR stakeholders.

Working with the United States Department of State and Office of Foreign Disaster Assistance, these teams provided vital search and rescue support for catastrophic incidences worldwide. In 1991, FEMA incorporated this concept into the Federal Response Plan (now the National Response Framework), sponsoring 25 national urban search and rescue task forces.

There are a number of models of deployment for ASAR teams. In the United States, with so many families owning animals, human SAR teams may include an animal technician for those situations that might include large numbers of dogs or evacuation of animals that might be fractious (embedded model). Or in situations where there are no ASAR teams, human SAR teams may receive "awareness" level animal handling training for addressing those evacuation situations where evacuees have pets that need additional control. In some states (e.g., Louisiana), animals that are brought in by the SAR teams are "handed off" to the local animal control/rescue when they reach the staging area. And the final model is a free-standing ASAR team that provides animal rescue—typically following human search and rescue efforts.

Many of the skills used in ASAR come from human SAR. The way that we organize the animal search process, the types of searches we conduct, and even the way we execute technical rescue look very similar to what is done for SAR. But there are a host of differences between human and animal behavior that may affect rescue success (Fig. 11.3).

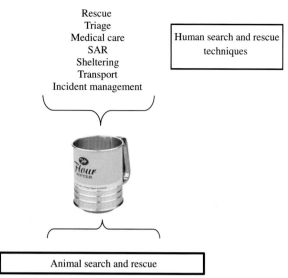

Rescue
Triage
Medical care
SAR
Sheltering
Transport
Incident management

Human search and rescue techniques

Animal search and rescue

FIG. 11.3   Animal SAR comprised of many best practices from human SAR.

Pictures of animals in distress pull at our heartstrings, and some folks experience a powerful urge to assist. These folks and groups are referred to as spontaneous (or self-deploying) uninvited volunteers (SUVs). Actually, there are numerous titles for groups that self-deploy, but in essence, if you decide to respond without a specific request from the AHJ, then you are considered self-deploying. Disaster brings thousands of SUVs to the response, and it creates challenges for the receiving group. Do you turn them away or try to absorb them into an already chaotic scene? And if that responder is really driven—which many are—and if the AHJ turns them away, they may drive to the next group. And before you know it, large enclaves of rogue groups are in the field.

Hopefully, there will never be another Katrina in the United States. Not only were the authorities dealing with response, but also they were dealing with thousands of responders looking for a way to help. Individuals were driving from every nook and cranny of the country—oftentimes driving straight through to help. Social media was showing photos of animals in horrible conditions and screaming for people to come help. When the AHJ tried to close the door on the SUVs, social media lit up on ways to get around the system, and rogue groups were clamoring for SUVs to come help them. It was chaos among the chaos. And it was incredibly unfortunate because these folks meant well; they made huge sacrifices to be part of the response mechanism, but the major groups did not know how to manage the onslaught of well-intentioned folks and supplies, and so, it took on a life of its own. Shelters began popping up all over the region. Animals were transported out of the state, doors were broken down and animals taken without a formal request.

Several good things and a few bad things resulted from this convergence. The rest of the states watched the situation closely and heard horror stories of SUVs and rogue groups and immediately began looking at their plans to see how they would prevent that from happening in their state. It became a difficult time for the reputable national groups to convince the states (and Feds) that not all animal welfare groups were rogue. Over time and with the experiences of Hurricane Sandy, the states and national groups recognized that the key to controlling the flow of SUVs is not to put up a barricade but to develop proactive messaging and just-in-time training and to work on ways to embrace and utilize volunteers rather than turn them away. And as to the rogue groups, they still exist, they still try to sneak into disaster areas, but communities and states are getting more savvy and controlling that much better today—thanks in large part to the lessons learned from Hurricane Katrina.

When Hurricane Gustav approached Louisiana in 2008 following a similar path as Katrina, the state and residents took notice. Gustav was the first storm to hit Louisiana since Katrina/Rita, and the state took advantage of those lessons learned including making evacuations more palatable for pet owners. Louisiana initiated the largest evacuation of the coastal parishes in its history and the movement of the families that needed transportation assistance went amazingly well. The Louisiana Department of Agriculture and Forestry (LDAF) quickly stood up their Incident Command Post (ICP) at their headquarters in Baton Rouge. The Louisiana State Animal Response Team (LSART) brought their leadership team to the same facility, and the coordination and collaboration between the two groups resulted in an extremely successful response with owners and pets staying together and coming home together!

Gustav provided an opportunity to work on another lesson learned from Katrina. One of the state and LSART partners was asked to provide a Subject Matter Expert (SME) to the ICP to manage the uninvited rogue groups. A number of rescue groups and individuals had started for New Orleans just as soon as they saw the storm coming. In their minds, this was going to be another Katrina, and they were bound and determined not to miss the opportunity to participate. The job of the SME was to follow various social websites and whenever there was an indication that a group was deploying, the SME would make contact with them and politely tell them that there were not any unmet needs, but if there was a need for assistance, they would let them know. That did not stop all of the groups, but it did stop most of the larger ones. Knowing that they would be called if needed eased their concern and for the most part controlled the influx of well-intentioned, self-deploying volunteers. The threat of being turned around at the state line may have helped as well.

ANIMAL SEARCH

Animals oftentimes do not want to be found and will in some cases do just about anything to avoid being found including climbing trees, wedging themselves into tight corners, or hiding deep into a culvert. Cats, especially feral cats, are much more inclined to hide when frightened. Domestic dogs can generally be coaxed out of their hiding spot especially for a tasty, smelly treat. A domestic cat may let out a meow when called and a dog may whimper or even bark, but both species will not generally assist in the extrication process when frightened. Humans will generally seek areas with a high likelihood of detection, whereas dogs and cats will opt for a tight, secure well-hidden spot until they are confident that it is safe to come out.

If dogs and cats are left in a home that begins to flood, they will continue to look for higher places to stay above water. Cats will typically be found on top of the highest cupboard—possibly in the space between the top of the cupboard and the ceiling. Dogs will swim until they can secure purchase to the next highest spot. Simply because of their size and inability to leap as well as a cat, they will have a tougher time surviving a flood in a home.

Both species can detect a hazardous situation, and during a fire or severe weather, they may be found under the bed, inside the lining of furniture, deep in the closet, or in crawl spaces. Consequently, when severe weather is looming, it is wise to kennel up your animal well before the storm.

Regardless of whether the victim has two or four legs, there are a number of basic principles that must be adhered to when working in the search and rescue field. First and foremost, the rescuer must not do anything that places himself or his team in harm's way. It's interesting that because animals are not able to communicate and humans are seen as caretakers for their pets, rescuers will oftentimes put themselves at risk to save an animal—much like we would see a parent put their own life on the line to save their child. It's human nature. Rescuers that can maintain a level head and not jump into a dangerous situation are seen as calloused or uncaring, but to lose a rescuer life to save a victim is still a lost life.

Starting the animal search and rescue efforts before human efforts are completed or under control may interrupt human rescue efforts and actually slow the animal response in the long run. This is especially true with water rescue. It may take 24–48 h to get a handle on humans needing water rescue. During that time, a large number of boats will be on the water along with helicopters performing airlifts. Having a well-intentioned animal rescuer in the middle of these activities places everyone at risk, may slow the human rescue efforts, and may result in a stop on animal rescue.

The amount of time to wait for animal rescue will depend on the type of disaster, the number of humans involved, the affected species, geopolitical and socioeconomic factors, and resource capabilities. Starting an animal rescue campaign before immediate human needs have been met may end up causing a good deal of frustration and even anger in the impacted community. When performing animal rescue in developing countries, it's common practice to take human and animal food when going out to the field. This type of approach is generally met more favorably and may even result in gaining the support of the community in locating animals and identifying additional resource support. Cross-training ASAR teams with human search and rescue teams will get the ASAR groups out earlier to not only do some great human rescue work but also gather some intel for the animal assessment process. During the Hurricane Floyd response, American Humane was asked if their water teams could be diverted to assist in human evacuations. They agreed immediately and quickly reconfigured their bots to make it easier to move people. When the families saw that animal rescuers were helping with evacuations, they were much more amenable to bringing their animals with them.

It has been mentioned previously that community leaders may prioritize the species that are most important to the community. That may have a significant impact on the search and rescue process. Large animal species are typically found in rural areas, they respond much differently than dogs and cats, and the rescue equipment is larger and more complex. Wildlife is most often found in rural and wooded areas; they are extremely elusive and dangerous to capture.

There are even differences with the rescue techniques for traditional farm animals. Sheep and goats respond (and herd) differently than pigs and fowl require nets and containment areas to trap, whereas cows are easier to gather but can be stubborn and obnoxious when asked to move. Consequently, the team composition and the equipment needs will be driven by the species needing rescue. It's not uncommon—especially in flood situations—where the large animal team finds a dog or cat that needs rescuing as well. Certainly, the ASAR team has the latitude to change direction occasionally as long as it does not keep them from their original mission. It is much easier for a large animal rescue team to pick up a dog than a small animal team to stop and rescue a horse. That is why having good communications in the field is so important and will be discussed later in this chapter.

In the early days of the response to Hurricane Floyd, the national response groups dealt exclusively with livestock. The emphasis quickly changed to pets as coastal communities became inundated with floodwaters. And like most flood situations, there were days where you were dealing with small and large animals. But regardless of the priority species, there is a similar prioritization process within species that takes place in the field. *It's the greatest number for the greater good.*

When doing animal rescue, the goal needs to be those animals at greatest risk and with the greatest likelihood of survival. This can be a very difficult concept to accept when you get into situations where there are large numbers of abandoned or feral animals. Most community members may insist that the domesticated, homed cat is more *important* than the feral cat when setting rescue objectives. That may be a tough pill to swallow for the animal welfarist. But the domestic cat, when rescued, will go back to a home that will care for it, provide it with medical attention, and bring joy to the family. The feral cat will be much more difficult to catch—thereby consuming a greater amount of time—and, once captured, either taken to high ground and released or held in captivity until the water recedes and then hopefully released. That cat will be extremely stressed and then held in an environment even more stressful than what they were "rescued" from.

There's no easy answer to this difficult situation. The chief elected official and the chief animal health official typically determine species prioritization. The incident commander and the command staff develop the objectives, and the ASAR team must follow those to the best of their ability—putting aside their personal thoughts or feelings—to complete the mission. Fortunately, even in a rigid command structure, there are opportunities to be heard, but if the team or individual members are not able to follow their directives, then they need to step down and go home.

As mentioned earlier in this chapter, during the early days of animal response for Hurricane Floyd, the emphasis was on large animals, and the very first request for American Humane was to help a pig farmer manage thousands of animals that have gotten out of the barns and were either swimming, stuck in trees, or dead and floating in the water. It was an absolutely horrible situation and one in which the rescuers struggled emotionally during the event and for years afterward. First, they were rescuing animals that few had worked with before, and second, within the population, the priority was for the adults that are very difficult to manage in a small johnboat. Had the priority been for the younger, smaller piglets, they could haul in six to eight of them on every sweep. And as all of that is happening, the water levels just kept coming up. More than likely, the team gave some serious thought about throwing in the towel and finding some other place to rescue dogs and cats, but they stayed with the mission.

After a difficult first day, AHA reached out to the community for help, and by six o'clock the next morning, an armada of boats were ready to come and help the farmer. The team leader gave the farmer a call to let him know they were coming only to be told that all of the animals had drowned. The team was devastated and angry. It wasn't for another 2 days that they discovered that their efforts of moving the pigs back into the barn very likely prevented a major contamination of the primary water source for the area. It is not the type of animal rescue the team had signed up for but a great example of how there may be underlying factors that drive an assignment. Would it have been *better* if the team had known the primary objective? Possibly. Would the team have stayed to complete their mission had they known? A good question to ask that team.

Fortunately, during floods, small animals and, specifically, domesticated pets are much easier to rescue than pigs. However, a large number of the small animals rescued will be strays. Approximately 15% of cats and dogs in a community are strays,[2] and some estimates are that 1/3 of the animals rescued during a flood are unowned dogs. Unfortunately, it's tough to determine that when rescuing the dog other than hints from their behavior, appearance, and the environment in which they were found. The bottom line in doing rescue work is as follows:

Place a priority on owned animals and address requests from owners before taking on strays or feral animals.

[2] Heath SE, Linnaberry RD. Challenges of managing animals in Disasters in the U.S. Animals 2015;5(2), 173–192.

SEARCH TECHNIQUES

Technology has greatly enhanced our ability to locate and rescue animals. GPS, GIS, heat-sensing equipment, powerful listening devices, and infrared night vision technology are just a few. The ASAR team may not have the latest hardware but armed with a basic knowledge of search procedures and an understanding of animal behavior, animals may be found that might otherwise have been missed.

The search team will work closely with the assessment team whenever possible. Depending on the type and scope of the disaster, the ASAR teams may be deployed before the assessment process is completed. This is more common for human rescue than for animal rescue given that in most cases, animal rescue will lag behind human rescue by 1–3 days. The assessment team will be able to provide the ASAR teams with much-needed information on access, high-risk areas, and animal impact. Incident command will consider that information when setting incident objectives that will then be passed down through the various levels of the operations section to the animal branch to determine tactics. An emphasis will be placed on areas and facilities that might house the greatest number of at-risk animals: animal shelters, zoos, aquariums, captive wildlife facilities, veterinary clinics/hospitals, research facilities, groomers, boarding facilities, breeders, and animal agriculture facilities to name a few.

After the scope of the disaster has been determined, a gridded map of the entire area will hopefully be available to the ASAR teams. Communities may be using United States National Grid (USNG) system or Military Grid Reference System (MGRS), and if those aren't available, their GIS group will likely have developed maps showing impact that the rescue teams can divide into search sections. Fig. 11.4 shows how one of the national groups organized their search and rescue efforts in New Orleans following Katrina. Unfortunately, very little information from the human assessment teams was reaching the animal search and rescue teams. Consequently, the ASAR teams were often frustrated and challenged by conditions that were changing quickly and areas either that they could easily reach

FIG. 11.4 Grid map generated by AHA for Hurricane Katrina. *Courtesy of American Humane. All rights reserved.*

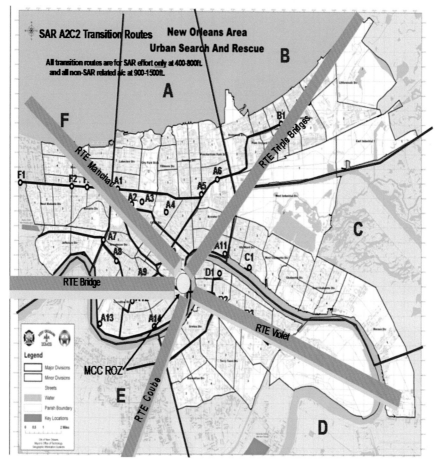

FIG. 11.5 New Orleans area urban SAR grids.

on one day but were inaccessible on the next day or where the water had completely receded. Consequently, the Animal Branch was dependent on feedback from the field teams each evening to determine search areas for the next day.

A closer look at the map will reveal latitude/longitude coordinates for homes where owners had requested their animals be rescued. The number of dots helped determine the size of the search area to ensure that teams were completing their sections in a timely way.

Fig. 11.5 shows how New Orleans has predetermined how the parish will be gridded into search areas. For communities that are prone to disasters, this saves valuable time and lets all of the response teams know the geographic area, the best route for approaching their area, and who will be in the adjacent zones.

Hurricane Rita (2005) followed right on the heels of Hurricane Katrina. The American Humane Association was tasked by LSART to support Calcasieu Parish. Immediately upon arriving, the NGO and Parish developed an assessment plan, and within hours of arriving, the assessment teams were in the field. Based on the findings from their assessment, a decision was made to feed in place rather than to bring in animals that were left in homes. Feeding in place options will be discussed in Chapter 12, but suffice to say that if that information had not been relayed accurately in their report, a large number of unsolicited resources would have shown up at the parish's doorstep to help rescue (Fig. 11.6).

There are three types of searches that may occur once search zones have been determined. A *hasty search* is typically a fast-paced visual inspection of the area accompanied by vocal or audio hailing. It's methodical and organized to ensure that the area is canvassed completely and yet quickly. The goal of the hasty search is to get a global picture of impact to determine potential number of animals at risk and resource needs. The search team may collect coordinates and determine priority areas for teams following behind. If the hasty search team finds an animal **at risk** that they can easily get out without taking them away too greatly from their mission, then they will deal with that animal and then move on. It's important to note that this is a quick process and the primary objective is not to rescue but to provide information on where to concentrate resources as they become available.

FIG. 11.6 Lake Charles, LA Search Grid, 2005. *Courtesy of American Humane. All rights reserved.*

FIG. 11.7 ASAR activities in Hurricane Floyd, 1999. *Courtesy of American Humane. All rights reserved.*

Fig. 11.7 is a photo of the hasty search team following Hurricane Floyd. This team did a quick boat tour around the mobile home park calling out for animals. When they heard sounds or saw obvious signs, they flagged the house. In addition, this team was also responsible for determining the search areas for the rescue teams that were gearing up at water's edge. This process may have taken 30–45′ but provided the structure and key information for the teams following to be more effective.

The *primary search* typically includes a walk-around of every building likely to contain animals—looking into windows and doors and calling out. In most cases, if resources allow, the primary search team does not enter the facility but will record for secondary team and move on. Finding a life-threatening situation would be the exception. It's critical that the search teams have determined a consistent way of identifying structures that might have animals at risk and of carefully identifying structures that might be compromised in some way to put responders at risk.

The *secondary search* consists of a thorough and systematic search of every room of every building within the impacted area. That may be room to room and layer to layer. Once again, the responder needs to be thinking of the most likely places where various species of animals are in the home. Animals that are caged such as reptiles, fish, and birds will likely have been moved upstairs in flood-prone areas; cats and dogs may work their way upstairs if access was available; but basic survival mode will drive animals upward in most situations. For nonflood situations, frightened animals will seek small, well-hidden areas such as deep into cupboards, inside the liners of furniture, on top of furniture, or in air ducts.

Carcass Removal

If a deceased animal is found, many jurisdictions will have the responder double bag the carcass; apply duct tape to secure it; and write the date, time, and responder information on the duct tape. If owner information is available, dispatch will want to call the owner to see if they would like the animal to stay on the property or go to animal control where it can be placed in the freezer. Typically, for small, outside animals, the carcass is bagged and disposed of in a method designated by the appropriate authority. Information is left at the property describing animals taken, date, and time and responder information. Large animal carcass removal is typically done by contractors as it may require a front-end loader or backhoe.

For a recent fire, a response team came across a property where an owner had evacuated and left his animals chained in the backyard. A situation found all too often in flood situations as well. Unfortunately, all of the animals perished in the fire. The responders documented the scene and working with animal control, were able to provide enough evidence to charge the owner with abandonment and neglect.

House Identification

Urban Search and Rescue teams have developed a system for marking houses that have been searched following a disaster. The photo below (Fig. 11.8) shows a home that was actually searched twice and lets emergency services see at a glance that the home was searched, when it was searched, by whom, and what they found. A single diagonal slash indicates that a search in the building is in progress. This is used to indicate searcher locations and to avoid duplication of the search effort. An X inside a square means "dangerous—do not enter!" An X with writing around it means "search completed," with the time (and the date if appropriate) written above the X, the team conducting the search written to the left side of the X, the results of the search (the number of victims removed, number of dead, and type of search such as primary or secondary) written below the X, and any additional information noted about the structure to the right of the X (Fig. 11.9).

These x-codes are used in a variety of situations and were prolific (and adopted and modified by other agencies) during post-Katrina operations.

There has been considerable discussion as to whether ASAR should try to include information within a US&R marking or have their own house marking system. The latter option was chosen given the large amount of information that might need to be conveyed in a house marking system related to animals. The NASAAEP ASAR Best Practice Working Group developed the following marking system to provide a method for informing SAR teams and emergency services of the animal status within or around a home.

ASAR uses a triangle rather than an X to distinguish animals from humans (Fig. 11.10). With the triangle pointing up, there will be four divisions with the bottom section being for identifying hazards, middle section for the date, second middle section for time, and top for agency conducting the search. These are preprinted forms on 5×7 or 8.5×11 with sticky back. Most of the agencies print their forms on very bright (neon) paper to be more recognizable

FIG. 11.8 US&R house marking. *Courtesy of American Humane. All rights reserved.*

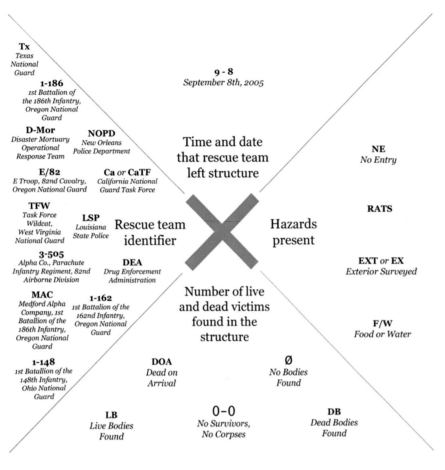

FIG. 11.9 US&R House Marking. *Courtesy of Wikimedia Commons/Georgelazenby.*

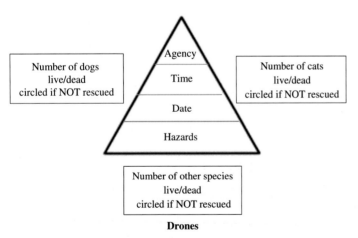

FIG. 11.10 ASAR recommended house marking. *Courtesy of NASAAEP ASAR Best Practice Working Group.*

from a distance. To the left of the triangle will be the status of dogs; to the right, status of cats; and below the triangle, status of other animals found in the house or on the property.

Drones

Drones, or, as the Federal Aviation Administration (FAA) prefers, unmanned aircraft system (UAS), or as the industry often refers to as unmanned aerial vehicles (UAVs) are already being used in humanitarian response around the world (Fig. 11.11). An unprecedented number of small and lightweight UAVs were launched in the Philippines after

FIG. 11.11   Drones, much like the one pictured here, were used to perform property inspections following 2017's Hurricane Harvey. *Courtesy of Pexels.*

Typhoon Haiyan in 2013. They were used in Haiti following Hurricane Sandy in 2012. And in 2014, they were flown in response to the massive flooding in the Balkans and the earthquake in China.[3] Drones have been a part of local SAR for some time delivering small rescue equipment much more quickly to victims in distress. Dropping a personal floatation device (PFD) to a drowning victim from a drone may take a couple of minutes—about the amount of time it takes to back a trailer down the boat ramp! Dropping a rope to a stranded hiker, a radio to SAR team, or emergency medical supplies to an inaccessible community are all examples of how drones can be used effectively to save lives.

Having a view from above provides an amazing advantage for defining the assessment process and for developing the ASAR strategy. Satellite imagery has been invaluable in recent disasters, but their use may not be available to the NGO, and the time needed to acquire images may limit their use. UAVs however are relatively inexpensive, provide higher-resolution images, and yield immediate feedback of the situation.

In the first 6 days following Hurricane Harvey, the Federal Aviation Administration issued more than 40 separate authorizations for emergency drone activities. The approvals were for assessing roadways and railroad tracks and the condition of water plants, oil refineries, and power lines. Total approvals exceeded 100 quickly, including some flights prohibited under routine circumstances.[4] The vast majority of the operations were conducted in conjunction with or on behalf of local, state, or federal agencies. Some applications were processed within hours, an unusually fast turnaround for federal safety regulators accustomed to days or weeks of analysis for such decisions.

The Red Cross utilized a drone for conducting assessments following Hurricane Harvey, "The measure of success for the American Red Cross on this pilot will be to prove that drones can help support, complement and accelerate the work already being done by our tremendous volunteers."[5]

Shortly after any major disaster, the skies are typically full of helicopters providing life-saving missions, and there is a concern that drone operators will interfere with rescue operations. The licensing and training of drone pilots and obtaining speedy access to segregated airspace are additional challenges. Following the El Dorado Hills Fire (2016), a drone grounded aircrews in certain areas causing additional risk for firefighters and citizens. The illegal use of the drone in the fire area resulted in the arrest of the drone owner.[6]

TEAM COMMUNICATIONS

The number one lesson learned following most disasters is *communications* or a failure thereof. In many cases, the cause of the criticism falls under three categories: interoperability, poor communication within a team, and the lack of effective communication equipment.

Interoperability refers to the ability of agencies to communicate. One of the primary lessons learned following 9/11 was the inability of fire/rescue to talk to law enforcement—they were on different frequencies, and there was not an audio bridge in place to rectify the problem. Poor communication refers to a team or agency that does not have

---

[3] https://www.virgin.com/virgin-unite/business-innovation/humanitarian-sky-drones-disaster-response.

[4] https://www.wsj.com/articles/drones-play-increasing-role-in-harvey-disaster-recovery-efforts-1504474194.

[5] https://www.reuters.com/article/us-storm-harvey-redcross-drones/red-cross-launches-first-u-s-drone-program-for-disasters-idUSKCN1BI2X9.

[6] http://www.latimes.com/local/lanow/la-me-ln-man-arrested-for-drone-in-fire-zone-20160715-snap-story.html.

protocols in place to direct the form, frequency, and type of communications needed in a disaster. This can also include communications between field and headquarters. And finally, the lack of effective communication devices can best be demonstrated following Katrina where all of the cell towers were either overwhelmed or nonfunctioning. The only form of communication available—and it was sketchy—was texting. None of the rescue groups at the time had satellite systems or backup forms of communication; fortunately, many have fixed that problem today.

Rescue teams need to test their equipment monthly and need to determine where communication gaps are located in their service area. The International Fund for Animal Welfare (IFAW) is based on Cape Cod, MA. There are a number of areas on the island where hills disrupt radio signals. IFAW conducted a full-scale exercise testing all of their equipment and included emergency management and amateur radio groups (Amateur Radio Emergency Service (ARES) and Radio Amateur Civil Emergency Service (RACES)) so as to identify gaps and develop solutions so that they would be able to effectively communicate anywhere on the Cape. The lesson learned from that exercise was if they positioned one of their trailers with a repeater at mid-island and changed the antennas on their vehicles, they were able to effectively communicate in about 95% of the area that included about a 40 mi circumference from their repeater. During their exercise, they also tested their satellite system, satellite phones, and low-frequency radios. Handheld radios are great for communicating within close proximity but if a team is going to communicate from the field to their base or even between teams in the field, they will need to invest in some additional equipment and training as discussed in the following section.

In emergency work, the success of an operation may hinge on effective communication. There are a number of forms of communication that can be used in search and rescue work:

- Voice. Vocal communication is more than using a radio; it's also knowing what needs to be said and when. In times of crisis, we need to recognize what needs to be discussed at the time and what can wait for when the dust settles. It is also the time for clear heads to prevail and for communications to be performed in as calmly a state as possible. That's assuming that you do not have helicopters above or standing next to a raging river. Consequently, there will be times when other forms of communication are needed—either when the surrounding noise drowns out communication or when you have a need for radio silence (approaching an animal).
  - When using vocal communication, use[7] the following:
    - Standardized language—not slang. Not every member of your team may understand local idiom.
    - Easily understood phrases. For example, people familiar to working with horses know that the "onside" or "near side" of a horse is the animal's left side. If a rescuer is instructed to approach the horse by the "near side," they will in all likelihood approach it from the side of the animal nearest to them, no matter if that was the left or right side of the horse. "Approach the horse from its left side" is a much clearer, more universally understood phrase.
    - Brevity—Don't use 30 words if you can use 10.
    - Repeat back—If you are not sure you are being understood or if you are not understanding, repeat back to each other what you thought was said. Get it straight.
- Radio communications. Handheld radios are the norm today for most rescue groups. Unfortunately, some of the groups are on UHF and others on VHF, and very few of the groups have access to audio bridges to be able to communicate among themselves:
  - Push to talk (PTT)—Most two-way radio sets make use of push-to-talk buttons. One of the most frequent criticisms of people new to PPT buttons is that they begin talking when—or before—the button is depressed. This causes the first few words to be cut off from the transmission.
  - Clear text. Do not use 10 codes or agency-specific codes. Clear, simple language.
  - Range. Most handheld radios are 4–5 W and might have a range—under perfect conditions—of several miles. In most cases, however, they will only suffice in immediate surroundings that means from one end of the fairgrounds to the other:
    - Mobile units (40 W) will have a bit more reach, but without a repeater in play, it's probable that none of the devices will be effective in communicating from the field to your base or camp.
    - The repeater is typically 40–50 W, but its value is in the ability to bring in a signal from a device and boost it along to another device, in essence multiplying the range of reception. If the repeater is strategically placed in the middle between field and base operations, you may find that communication between groups is now possible.

[7] Adapted from IFAW Slackwater Rescue Course—Nicholas Gilman and Dick Green.

- Inappropriate language—All radio usage is subject to regulations put forth by the Federal Communications Commission (FCC).[8] Professional-level two-way radios are directly licensed through the FCC, while citizen's band radios are regulated by the FCC but require no license. In either case, FCC regulations—which apply to all two-way radio use in the United States—forbid the use of inappropriate or offensive language. Use two-way radios professionally and with regard to everyone who may be on that channel.
  - Batteries and power sources—If your rescue work is going to rely heavily on the use of two-way radios for communication, take extra care to ensure that extra batteries, chargers, or other means of powering the radios are in place,
  - Prepare for no radios—Radios are very handy, but complete reliance on them for communication is unrealistic. Practice other means of communication so that a lack of radio communication will not cripple your response.
- Nonvocal communication. There will be times when it is preferable to use whistle blasts or hand signals over radios or simply shouting.
  - Whistle commands—Unlike radios, whistles are inexpensive, can be used without hands and or battery power, and are far more reliable. They are not as communicative as spoken language, and their range is limited to approximately 100 yards. Whistle commands are standardized in inland water rescue circles:
    - One blast—"Look at me!" "Attention!" "Stop!"
    - Two blasts—"Up," "go up," "look up," "haul up," "upstream," "come upstream," etc.
    - Three blasts—"Down," "go down," "look down," "rope down," "downstream," "come downstream," etc.
    - Three blasts repeated—"Emergency!" (these whistle blasts should be repeated over and over again.)
  - Arm and hand signals are as follows:
    - Either hand tapping top of head—"I'm okay!" "I'm ready."
    - One arm raised up in air and left there—"I need help" (arm may be waving or still).
    - One hand tapping top of head, while other arm points at someone—"Are you okay?" "Are you ready?"
    - Either arm in chopping motion in a given direction—"I want to go that way." "You go that way." "It's over there."
    - Both arms crossed in front of the body—"Stop."

ANIMAL RESCUE

Animal rescue equipment has become much more sophisticated with harnesses, glides, sedation and capture equipment, and even our ability to communicate with high-range radios, repeaters, and satellite devices. All of these advances and technology come with a price and possibly even a loss of effectiveness. In the recent floods in Louisiana (Sep. 2016), there were a number of rescue groups that showed up with airboats and V-hulled rescue boats and were unable to access the most critical areas. Those that had the smaller (14–16′) johnboats and smaller engines were pulling out animals quickly and safely. As the larger boats were looking for adequate depth and access areas for their trailers, the groups with the smaller rigs were literally carrying the boats and equipment to the water's edge. Granted, the larger boats are great in deep water, but an important rule in animal rescue is as follows:

Keep it simple and safe (KISS).

This is true with equipment as well. A catchpole, net, and leash will take care of nearly every situation you come up against when rescuing dogs and cats. Those three pieces of equipment are light, take up little space, and are relatively inexpensive. Following Hurricane Matthew (October 2016), the ASPCA's field teams utilized 14–16′ johnboats with 2–3 responders per boat along with 3–4 plastic crates and armed with a catchpole, net, and leash and rescued scores of animals daily for over a week.

Keep it simple and safe.

As mentioned earlier, humans for the most part will assist in their rescue. They will shout out, reach for a line, and even swim to a boat to be rescued. A friendly dog may welcome a rescuer in a red dry suit, helmet, and PFD, but the rest of his friends and certainly most cats will not be excited to see such a strange and intimidating figure. And in many

[8] Federal Communications Commission.

FIG. 11.12   Villahermosa, MX 2008. *Courtesy of The International Fund for Animal Welfare. All rights reserved.*

FIG. 11.13   Hurricane Ondoy, 2009. *Courtesy of The International Fund for Animal Welfare. All rights reserved.*

cases, they may inflict great harm when you attempt to rescue them. Even the most domestic loving cat can become extremely difficult to handle in time of distress and will even strike out at their owners.

Injured animals are even more defensive.

While working the floods in Tabasco, MX, in 2008, a black lab rottie mix was found standing in chest deep water not making a sound and making no move toward swimming away or coming toward the boat (Fig. 11.12). The boat approached the dog, and the rescuer reached down and, using a scruffing method, lifted the dog into the boat. As the dog was coming up over the gunwales, the rescuer reached back to lift its rear legs over, and the dog swung its head and bit the rescuer. The rescuer had let his guard down slightly as there was no indication that the dog was fractious until he grabbed his broken leg.

Cat bites can be extremely painful and dangerous as their sharp teeth quickly find bone and are surrounded by bacteria that can cause infection if not properly attended to. Consequently, it's often best (and easier) to use a net when catching cats in times of emergency. A good cat handler will be able to get a quick scruff and round up the back legs, but small cats and strong cats with little fur around their neck (feral tomcats) can be difficult to control. And when they are capable of squirming, they will lash out with their claws and sink their teeth if the opportunity presents itself. If a net is not available, then the action of capturing and putting into a carrier must be very quick and requires a good deal of experience.

In 2009, IFAW responded to the flooding in the Philippines following Typhoon Ondoy. As the team was doing their assessment, they came across the roof of a garage that was still above water, and there were three young cats hiding in the rafters extremely hungry and dehydrated—and not overly friendly (Fig. 11.13). The team was able to lure them out to get to some food, and when enough of their body was exposed, in one smooth motion, the rescuer scruffed them around their neck and launched them to the net person who caught them in midair and was able to then put them into a crate. That team had done that maneuver enough times where their actions were anticipated and both animal and rescuer were unharmed. Transferring cats from nets to a crate can be tricky and will require coordination and quick hands!

ANIMAL SEARCH AND RESCUE: LOCATE, CAPTURE, CARE, AND RE-UNITE

A typical ASAR mission begins with the dispatch or *request for rescue* (Appendix I). The ASAR team has either the jurisdictional responsibility or statutory authority to rescue an animal from a residence. Whenever possible, the requestor will provide a key to the ASAR team so that a forcible entry is not necessary. In the case of an ASAR team working under the delegation of the Authority Having Jurisdiction, it's always preferable to have owner's written permission to break and enter. Following Hurricane Katrina, the recognized rescue groups received well over a thousand requests to rescue abandoned animals. Some of those requestors either mailed or dropped off their keys, and many others signed a *break and enter release form*.

For those situations where an animal must be removed from the premise and a key has not been provided, the ASAR teams must be trained in methods for entering a secured residence (Fig. 11.14). The key is to make every attempt to enter the premises in such a way that you can leave it secured when you leave:

- Look for possible hiding places that the owner might have left a key.
- Check all windows and doors. Oftentimes in the summer, residents will leave their window open with a screen in place. If the tabs for the screen have been placed to the outside, it may be possible to pop the screen for entry.
- Pulling air conditioning units is generally an easy process and can be resecured when you leave.
- Use upper windows if none of the above work that will detract looters from a similar entry. If you must break a glass, break just enough to access the lock and make sure to tape up any damaged glass in case other animals are still in the home.
- The last resort is breaking down the door and quite frankly, an option that hopefully is never used.

Following Hurricane Ike, there was a "rescue" group that took great pride in breaking down doors to rescue animals. In most cases, they had no knowledge that the animal was in distress or if the owner had actually

FIG. 11.14 ASAR team gaining entrance to home during Katrina response, 2005. *Courtesy of American Humane. All rights reserved.*

evacuated. They were simply using the opportunity to videotape their heroics. That group's leader was escorted out of the state but his actions cast a cloud over all of the other rescue groups and to this date, there is still friction between state emergency management and the animal rescue community. There were so many similar examples following Hurricane Katrina where individuals/groups used the disaster as an opportunity to break down doors, pulling out animals not in need and even snatching animals out of backyards when the owner was in the house. ASAR teams can be effective without acting like a SWAT team and rescue animals in need without sacrificing the integrity of the home.

Keep it simple and safe, don't lose track of the mission objectives, and rescue ONLY what needs rescuing.

UNIQUE FEATURES OF ASAR

Whereas humans can provide their address and phone number to the rescuing team, abandoned or rescued pets need to be identified at their point of pickup. This is the start of a *chain of custody*, and the rescuer needs to take ample time to properly identify the location and condition of the animal so as to improve their chances for rehoming. There are a number of electronic and wireless options available to the ASAR teams but at a minimum; address and/or GPS coordinates, date, time, description of animal, and name of rescuer should be recorded when the animal is rescued. That information needs to follow the animal throughout its time away from its owner. If electronic copies are not available, a field card such as the one that follows should be used (Fig. 11.15). When the animal is placed in a transport carrier, that information should be attached (or duct-taped) to the carrier and eventually used to develop an intake card at the shelter. Triplicate forms may assist in this process and provide the needed redundancy to ensure proper chain of custody.

ASPCA

Responder _____ ID Number _____

Date/Time _____ (Dept- year-case#-initials-#) (D-2013-03-DG-01)

Address/GPS coordinates _____

Contact info to reclaim animal

_____

Animal info: _____ _____ _____ _____

Species   Breed   Description/ID   Condition

FIG. 11.15 ASPCA field card.

The transport of animals has many of the same requirements as for humans and was addressed in Chapter 8. Unlike human transport, animals need to be fully contained and in carriers that allow for proper ventilation. And whereas humans can let you know if they are hungry, thirsty, or need to go to the bathroom, animals will be dependent on the rescuer to provide for them and anticipate their needs. When transporting animals from the field, it's common to use large trucks such as National Guard vehicles, and as long as the animal has proper ventilation and is secured in their kennel, for the most part, this is an acceptable form of emergency transport.

HURRICANE RESPONSE

Hurricanes can bring about a wide range of response efforts from dealing with life-threatening winds to devastating floods, and those scenarios often require an army of well-trained individuals and a host of equipment and supplies. Fortunately, weather monitoring agencies are capable of providing something that we didn't have 50 years ago—time to prepare. There will always be a degree of uncertainty when it comes to forecasting, but emergency management—

especially in the coastal states—has invested significant funds and research into the art and science of hurricane forecasting.

If there was a good thing about tropical systems, it's that they travel rather slowly, which means that a community may have days to prepare and evacuate if needed. It also means that the response community will have ample time to prestage resources to shorten their response time. Given the degree of uncertainty with every storm, pinpointing where a storm will hit and where a team may want to prestage resources can be tricky. Recent storms like Harvey and Irma (2017) and Michael (2018) were a bit more predictable, so teams could lean forward and predeploy resources to neighboring states. For Harvey, the ASPCA staged in Lake Charles, LA, and for Irma in Duncan, SC. These sites were far enough away to keep the team and equipment out of harm's way but close enough that their response time was measured in hours rather than days. Prior to the arrival of Hurricane Matthew (2016), the ASPCA and Code 3 Associates staged in Waycross, Georgia—close enough to Florida and several hours from South Carolina that was the projected path.

Animal Evacuation and Relocation

Local government typically imposes a minimum holding time (typically 3–5 days) for animals that are brought into a community shelter. After their holding time is complete, those animals are available for adoption. When a disaster is imminent, shelters try to move these unclaimed animals out of the shelter to ensure their safety and to make room in case more animals arrive after the disaster strikes. This process of moving unclaimed animals is referred to as "relocation" or "relo."

The goal in a relocation effort is to place the at-risk animals in parts of the country where there is the greatest likelihood for finding new homes. A number of the NGOs maintain a large network of receiving shelters for this purpose. In 2017, the ASPCA Animal Relocation team helped transport more than 1600 homeless animals out of communities struck by disasters.[9]

Relocation efforts played a critical role before and after Hurricane Irma made landfall in September 2017. Given the general agreement on the tracking of the storm—and with memories of Matthew fresh on residents' minds—evacuations were underway well before Irma even hit Cuba. Animal shelters in Florida, Georgia, and South Carolina were seeking assistance in moving their unclaimed animals out of state. It is important to note that shelters in the south move pets to areas in the north all year long and many shelters have established partners that are willing to take their unclaimed pets and, in some cases, are even willing to come pick them up.

In a disaster, with limited resources, a rescue group cannot afford to take a transport vehicle out of service for any length of time. In the case of Irma, the ASPCA set up a way station in Duncan, South Carolina. This 40,000 sq ft emergency shelter temporarily housed evacuated animals from three states and scores of agencies. Nearly 600 animals were sheltered there until the Animal Relocation team identified receiving shelters.

In the continental United States, it is relatively easy for a family to evacuate with pets, assuming plans were made early and transportation is available. However, living on an island is more challenging for evacuating with pets. Seats are limited, airlines have become more restrictive on how pets fly, and it is expensive. Consequently, a much smaller percent of residents evacuate from an island with their pets than do their counterparts on the mainland. In the case of Hurricane Maria hitting Puerto Rico and the Virgin Islands in September 2017, hundreds—if not thousands—of animals were displaced, abandoned, or free roaming.

Although evacuating an animal by air is faster, it is not as cost-effective as by land. For example, an air transport of 100 animals could easily cost up to $50,000. In Puerto Rico and the Virgin Islands, it was difficult to get in larger airplanes, which meant that multiple trips were needed using smaller planes with a capacity of 30+ animals. Nearly 400 cats and dogs—either not owned before Hurricane Maria or surrendered afterward—were relocated from St. Croix to the mainland.[10] According to the Humane Society of the United States, over 2000 animals were airlifted from Puerto Rico.[11]

[9] Green RC. Animal relocation efforts during Hurricanes Harvey, Irma, and Maria. DomPrep Journal Feb. 2018;14(1):16–18.

[10] Green RC. Animal relocation efforts during Hurricanes Harvey, Irma, and Maria. DomPrep Journal Feb. 2018;14(1):16–18.

[11] http://www.humanesociety.org/news/press_releases/2017/11/the-humane-society-of-the-112717.html?credit=web_id80597821.

WATER RESCUE

The water environment creates a number of challenges for the ASAR responder. Water conditions in a flood can oftentimes be nasty and full of contaminates (Fig. 11.16). "Plucking" animals from the water and safely bringing them into the boat require skill, strength, training, and proper equipment; simply finding enough trained responders and appropriate watercraft can be a challenge.

FIG. 11.16   Contaminated water in Louisiana floods 2016. *Courtesy of The American Society for the Prevention of Cruelty to Animals. All rights reserved.*

Water rescue can be classified in two ways: slack water, slack or slow-moving conditions, and swift water, moving water conditions. Unfortunately, slack-water conditions can have characteristics of the swift water environment, and one must always be aware of the potential for changing and mixed conditions. Even in floods around fields and lakes, there can be areas where the water moves quickly. There are a number of slack-water courses being taught for animal first responders (ASPCA, IFAW, Code 3, and AH), and they are all geared to provide the basic knowledge to assess whether the rescue of an animal in a water environment can be accomplished:

safely and responsibly with the equipment and personnel on hand.

Animal first responders must have a profound respect for the dangers inherent in conducting rescues on water. They need to learn how to assess water conditions and to design rescue options from the simple to the complex. They need to know how to use equipment to make their efforts safer and more effective. And most importantly, they will have an understanding that

personal and team safety is their number one priority.

Basic and technical skills presented in the various courses must be practiced and oftentimes adapted to fit team structure, capabilities, and water situations. The certificate of completion for the courses ranges from 3 to 5 years, but if the first responder does not practice the skills at least quarterly, memory and proficiency will fade quickly. The 16 h spent in the course is just the beginning to a commitment to practice and complete additional trainings.

Water Rescue Equipment

There are a number of boat and engine configurations used in animal search and rescue and are based on a number of factors including water conditions, training, rescue equipment, hauling capacity, trailering capability, and budget. Human rescue teams typically opt for larger (greater than 16′) boats that will accommodate six or more victims and designed to reach them quickly. These larger craft provide opportunity to enter into more challenging water conditions and to provide additional seating for evacuees. But they also restrict access into shallower waters and in some cases are difficult to launch without a ramp or sloped surface.

Animal rescue teams typically do not transport large numbers of animals, and they are not as concerned about speed, so they generally opt for smaller boats that maximize stability and floor space for stacking crates and cages and a style that puts the edges closer to the water (shallow draft) to ease bringing animals into the boat.

For the majority of the NGOs active in water rescue, the most commonly used boat is a 14–16′ flat-bottomed johnboat with a 10–25 hp engine. In slack-water rescue situations, there will be times when equipment needs to be portaged or walked through shallow water (Fig. 11.17). Areas that had 20′ of water on 1 day can be down to a foot or two the next, so craft that are easy to launch, carry, and navigate through and around tight spaces are the favorites among the animal rescue groups.

FIG. 11.17 Portaging equipment and animals. *Courtesy of The American Society for the Prevention of Cruelty to Animals. All rights reserved.*

Boats are typically classified by construction or by shape. For example, boats composed of aluminum or steel will have "rigid" hulls versus inflatables that typically have large pontoons. Some inflatables will have a rigid bottom or hull to better protect the fabric and are referred to as semirigid hulls.

Aluminum boats are very popular among rescue groups and fishermen because of their portability, durability, and cost. Aluminum boats can be welded or riveted. Riveted johnboats are extremely light and therefore ideal for areas when portaging gear but more susceptible to damage from wear and tear. Welded johnboats are heavier but much more durable and rated for larger engines. Johnboats come with flat bottoms or V-hulls. As mentioned above, the flat-bottomed boat is popular with many animal rescue teams as it offers more usable floor space but is not as stable in open water as a V-hulled boat.

Johnboats are easy to trailer and many groups will stack multiple boats on a single trailer when teams are working in close proximity of each other or when access is restricted—another advantage of a johnboat over larger craft. Since a

FIG. 11.18 Multiple boats on a single trailer. *Courtesy of American Humane. All rights reserved.*

FIG. 11.19   Inflatable boat. *Courtesy of Inmar Inflatable Boats.*

number of groups (AHA, Code 3, IFAW, and ASPCA) have large equipment trailers, it's possible to get 3–4 johnboats on a single trailer in the back of a 36′ trailer (Fig. 11.18).

The size of engine used will be determined by the boat rating, water conditions, portability, and personal preference. The rating plate (Fig. 11.19) can be found at the back of the boat typically on the transom near where the engine attaches. The rating plate listed here is for a 14′ johnboat rated for three people or 480 lb and for a 15 hp engine. Certainly, a smaller engine could be used and in some cases warranted when the goal is to keep weight down and when you know that you will be in true slack-water conditions.

Some rescue teams utilize electric trolling motors that work well when speed and water conditions are not an issue. The trolling motor also runs quietly that can be advantageous when working around frightened animals. They are very light including a deep cell battery and can be positioned at either end of the boat if necessary.

A number of rescue groups continue to utilize inflatable boats for rescue operations. The primary disadvantage of the inflatable boat is its susceptibility to damage. In flood work, you will oftentimes be in waters riddled with debris and obstructions and that can be tough even on the most durable fabric. Military and rescue-grade materials have made current generation inflatables incredibly durable and portable. In many cases, the boats can be loaded into the back end of an SUV and inflated at the site, eliminating the need for getting trailers into difficult spots. There are a number of inflation systems that can fully inflate a 14′ boat in less than a minute with a single tank of compressed air.

Inflatables draw very little water and can get into spots that challenge even the smaller johnboats. They can handle the roughest of waters and have been the mainstay for the United States Coast Guard (USCG) rescue teams for years.

High-end inflatables are very pricey and, with floorboards, very heavy but incredibly stable which is a major advantage when moving responders and animal around in the boat. They require a longer-shafted engine and more horsepower and maintenance is typically higher than for rigid-hulled boats. Bottom line is that inflatables are a valuable piece of rescue equipment for a number of conditions and situations.

Airboats or fanboats are a flat-bottomed vessel (johnboat) propelled by an aircraft-type propeller and powered by an either aircraft or automotive engine (Fig. 11.20). They are commonly used for fishing, hunting, and ecotourism but have become a staple in some rescue situations. Very common in the south and southeast, airboats were a common sight following Hurricane Harvey (Aug. 2017) when the Cajun Navy came to Texas to assist. They work extremely well in marshy and/or shallow areas where a standard inboard or outboard engine with a submerged propeller would be impractical. The biggest disadvantages with the airboat are noise and weight. They are extremely loud, and the large motor on the back makes it impossible to portage and therefore will need an area to launch from a trailer. Airboats can also be extremely useful when conducting initial size-up of a flooded area. Following Cyclone Ketsana (2009), IFAW utilized an airboat to quickly assess the situation for three different villages in less than an hour, a task that would have taken most of the day by johnboat—or multiple days by banca!

In situations where there is mud and shallow water, a mud motor or long-tailed motor can be extremely effective (Fig. 11.21). These motors as seen here require very little water to navigate and with their shaft and propeller position

FIG. 11.20   Airboat. *Courtesy of The American Society for the Prevention of Cruelty to Animals. All rights reserved.*

FIG. 11.21   Gator tail engine.

are perfect in situations where water conditions are unknown or where there is significant plant growth and mud. During the Hurricane Harvey response, these units were getting into areas that other boats were unable to access. They are heavy (the engine shown here is over 250 lb) and expensive ($8000), but if you have the budget (about $20,000 complete) and the need, this configuration works great in a number of water situations.

Regardless of the type of boat and engine used, proper maintenance will ensure their reliability, state of readiness, and longevity. Even the larger NGOs do not get their equipment out the door on a regular basis. On a particularly busy year, teams may respond three to six times. That means that equipment sits for lengths of time and that can be hard on equipment. Most rescue teams inspect their equipment that includes running engines at least weekly or monthly.

An example might be helpful to bring this point home. Recently, an NGO supported a partner agency to respond to the worst flooding in their history. Several years earlier, the NGO had granted their partners the funds to purchase a 16′ inflatable boat and 25 hp engine. When the NGO arrived, the partner agency pulled the boat out of the bag to find that the mice had left little intact—the $5000 boat was ruined and the motor had not been started for 2 years.

Flooding presents a host of hazards including household, industrial, and agricultural contaminants and sewage and treatment plant waste (Fig. 11.22). Consequently, all water-based rescuers should be in PPE that will prevent potential contaminants from coming into contact with the skin.

FIRE RESCUE

The role of the animal first responder in the initial phase of fire rescue is primarily to facilitate evacuations. Animals react similarly to humans when placed in a life-threatening situation—they attempt to run. Trying to evacuate an animal when fire is knocking on the back door is a difficult process even for the most seasoned responder. But that is exactly what might be asked of the animal first responder when a large fire is approaching and the owner was not able to evacuate their animals.

The second phase for animal rescue during a fire response is search and rescue. Typically, animal control will receive numerous requests from residents asking for assistance in checking on their animals. If animal control is recognized as a member of emergency services, they will be allowed access into the burned areas. If the animals had an opportunity to run from the fire, there is always a chance that they might have survived. Dogs are fairly mobile and will run great distances as needed, whereas cats typically look for a place to hide out such as in a culvert or pipe (Fig. 11.23).

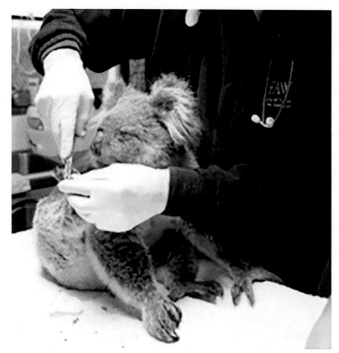

FIG. 11.24   Treating burned koalas after Australia fire, 2009. *Courtesy of The International Fund for Animal Welfare. All rights reserved.*

Wildlife can survive a fire as they are mobile. Slower-moving animals will oftentimes succumb to the heat or smoke, but members of the deer family may survive. Following the Black Saturday fire—the largest wildland fire in Australia's history—IFAW mobilized a veterinary team (Fig. 11.24) to treat the various wildlife that were able to survive. Treating wild animals is tough at best, but treating wild animals that have been severely burned is extremely difficult, and in many cases, the only humane treatment is euthanasia.

Animal rescuers working in fire areas will also want to carry items that will assist in their evacuation support such as flashlight, GPS, cell phone, toolkit, flares and reflective triangles, first aid kit, and animal capture equipment. Quite possibly, the most important pieces of equipment will be a truck and trailer. Training curriculum for animal first responders as it relates to fires should include fire behavior, vehicle and trailer operation, trailer loading techniques, animal tracking and intake, and animal first aid and triage.

Transport Vehicles. Large animal evacuations are the most often requested resource the animal rescue groups receive with a fire. Small animals are much easier to evacuate than livestock, and for the most part, families do a good job of getting their pets out. Livestock on the other hand require specialized equipment and additional preparation time to evacuate. And unfortunately, many households located in fire-prone areas do not have an evacuation plan, do not have access to trailers, and have not practiced trailering with their livestock so that when a fire is approaching their back door, their only choice is to open the gates and let them run.

All vehicles used in disaster response must be well maintained and functioning properly. Many local evacuation teams utilize their personal trucks and trailers, and the expectations on the farm are much different than in the danger zone. You can imagine the frustration and danger associated with a broken-down vehicle in the middle of the only access path. Emergency service vehicles are trying to get in, folks are trying to get out, and a rescue group's vehicle is obstructing traffic. This is putting human and animal lives at risk—and it will make it difficult to get an invite back on the next disaster.

The most common vehicle used in evacuation work is a ¾ or 1 t 4WD pickup. The four-wheel drive is critical as firefighters may need to reroute the evacuation team off-road to access properties. Steep and dangerous terrain will take its toll on an older vehicle and worn tires. You never know when you will find a fueling station, so always keep your tanks full and carry extra fuel if necessary. Extra spare tires and a toolkit with the basics needed for minor repairs are essential.

A smaller horse or stock trailer is preferred for emergency evacuation work (Fig. 11.25). There will be situations where the larger trailers may have access, but for the most part, a 2–4-horse-capacity trailer is a good compromise between size and accessibility.

TRAILERING

A number of local and state rescue groups have developed courses on trailering a horse as it is not a skill you will learn from a book or online. It takes a good deal of practice and confidence on both ends of the lead rope to be effective. And remember that even the best of horses may develop a change of plan when fires are licking at their hooves (Fig. 11.26).

There may be situations with some species where fear and the lack of practice will win over loading and that animal may have to be left behind to save the remaining animals. With that in mind, some teams prefer to trailer the easiest ones first so that they aren't held up by the difficult horse. In many cases, there will be a leader among the herd, and with a little luck, the difficult ones will follow their lead. The following tips may assist with the more difficult animals[12]:

1. Have the horse approach the trailer calmly until they reach the point when they are uncomfortable and want to stop. Let them stand there for a moment and look around. Then, lead them away from the trailer and repeat the process. Each time they should be able to approach the trailer a little bit closer. Wait for when they start to want to investigate and touch the trailer with their nose and accept and reward the slightest try.
2. Turn on the interior trailer lights and open the windows if available.
3. A stubborn horse draws a crowd. Keep the noise down and the two-legged supporters to a minimum. There should only be one voice, and that is from the team lead, and it should be calm and soft.

---

[12] https://www.doubledtrailers.com/seven-tips-for-safe-horse-loading/; https://www.doubledtrailers.com/5-tips-from-a-parelli-trainer-on-how-to-load-a-difficult-horse/.

4. Have familiar smells in the trailer—horse blankets, bedding, etc.
5. Some horses respond best to requests to move forward from in front and some from behind.
6. Incentives may work, and tranquilizers may be needed.
7. Wait to shut the door (if possible) until everyone settles down.

ANIMAL TRACKING AND ACCOUNTABILITY

When the animal evacuation team arrives on the scene, depending on protocol, they may either call in their arrival or make note of the time on their dispatch sheet. If the owner is there and they requested the evacuation, then a quick review of what is to be taken and ensuring the request matches the transport number is the first step. If the owner did not make the request (came in from concerned party), then the owner will need to sign a release or contact the AHJ to approve the evacuation. If the owner does not want the animals taken, then that becomes an issue between the AHJ (or law enforcement) and the owner. If you have a long list of animals waiting, do not get hung up for any length of time fighting a legal battle. Let dispatch take care of that and move on down your list.

When animals have been identified for transport, a transport log or inventory sheet should be completed to allow for tracking animals. These forms should include date, time, address, and number and disposition of the animals (Appendix J). If you are loading 12 pigs and the fire is approaching, then load 12 pigs and put on the inventory sheet, "12 pigs." But anything you can do to ensure that every animal you pick up from a property is properly identified will go a long way toward a happy reunion. If you have time, take a photo of the animals and create an animal identification. If you are being dispatched by animal control, this might already be built into the dispatch process through their case numbering system. If no numbering system has been assigned, ask the unit or branch lead for direction. If all else fails, use the event name and owner's name and number by species (see Fig. 11.15). If you keep your eye on the ultimate goal of reuniting animals with their families, the tracking and accountability process will take on greater importance (Fig. 11.27).

FIG. 11.27   Rescued cat from Lake County Fire 2015. *Courtesy of The American Society for the Prevention of Cruelty to Animals. All rights reserved.*

During Hurricane Harvey (Aug. 2017), the Cajun Navy appeared en masse into the flooded areas of Texas. Hundreds of boats and well-intentioned responders arrived and immediately started rescuing people and their pets. Undoubtedly, the Cajun Navy saved lives. They also created some real challenges for the various animal welfare groups that were on the receiving end. Many of the animals arrived without paperwork and in a number of cases were separated from the families making reunification difficult. There is a time when your only choice is to "pluck and run," but in most cases, the animal evacuation team should use at least a rudimentary recoding system to enhance the likelihood of an animal being reunited with its family.

ANIMAL TRAPPING

Invariably, there will be dogs and cats that are unwilling or unable to come to a rescuer for help when called or simply are too afraid to approach a rescuer. When their natural habitat has been destroyed by a disaster, the animal may linger around the impacted area for some time in search for food and safety. Following the Tubbs Fire (2017),

FIG. 11.28   Cat trapping following Joplin Tornado (2011). *Courtesy of The American Society for the Prevention of Cruelty to Animals. All rights reserved.*

FIG. 11.29   Cat Trap. *Courtesy Pixabay.*

Sonoma County Animal Services received numerous calls requesting support finding and capturing less than friendly pets. The most common form of capture is the live trap that typically provides one or two doors and a platform that will drop the door when an animal steps in for the food. During the Joplin response (2011), the ASPCA and the Joplin Humane Society trapped 155 animals. Feral animals were treated, identified, and released, and owned or "community" animals were reunited or available for adoption after the requisite holding time (Fig. 11.28).

There are a host of devices available, but animals will likely enter any device if it is placed in the setting and has the right scent. Many groups will suggest that prior to trapping, you establish feeding stations that will draw an animal into a specific area. If the intent is to eventually trap, then save the good smelly canned food for trapping days. Early morning or evening seems to work well, but regardless of the time, never leave a trap unattended for any length of time and don't leave a trap out at night. There are a number of websites that provide excellent guidance on how to set traps and transport and transfer animals (Fig. 11.29).[13]

FIRST RESPONDER SAFETY

Animal first responders will not be fighting fire, but they need an awareness-level understanding of the conditions in which they might be asked to work in. It is recommended that all animal evacuation team members complete S-190:

[13] https://www.aspcapro.org/sites/default/files/TNR%20workshop%20handbook%20printable%20final-4th%20printing.pdf.
[14] https://onlinetraining.nwcg.gov/node/169.

Introduction to Wildland Fire Behavior.[14] This is an excellent awareness-level course offered online by the National Wildfire Coordinating Group.

Wildland fires get started from a host of causes from unattended campfires to lightning. But the number one cause is human error. As much as 90% of wildland fires in the United States are human-caused[15] ranging from cigarettes, slash burning, and arson. Regardless of the cause, fires can be traced to the starting point—or *point of origin*. This is where some form of ignition source (match, lightning, etc.) comes into contact with a fuel source.

As fire spreads, typically, one side will burn more quickly, and that is referred to as the *head* of the fire. The *flank* of the fire is the perimeter of the fire that is roughly parallel to the main direction of spread. And the *rear* refers to that part of the fire spreading directly into the wind or downslope (opposite of the *head*). The fire perimeter is the outer edge of the fire, and frequently, there will be fingers of fire that project from the main body of the fire. Occasionally, there will be *pockets* or *islands* of terrain that remain unburned as the fire progresses, and sparks from the main fire may cause *spot fires*. The safest

FIG. 11.30 ASAR team working in the burned area. Lake County Fire 2015. *Courtesy of The American Society for the Prevention of Cruelty to Animals. All rights reserved.*

place for ASAR responders is in the "black" or areas where the fire has already burned through and there is no active fire in the area (Fig. 11.30).

Fire can also be described by the level of active burning. A *smoldering* fire is burning without flame and barely spreading. A *creeping* fire is burning with a low flame and spreading slowly, whereas a *running* fire is spreading rapidly with a well-defined head. When fires get very active and the wind kicks up, *spotting* may occur where sparks or embers carried by the wind start new fires. As the fire spreads across the vegetation and comes to a tree, it may work its way up the tree that is referred to as *torching*, whereas *crown* fires advance from treetop to treetop. Fires seldom burn at a constant rate and will experience *flare-ups* or sudden acceleration in the rate of spread. Fires will often be so intense that they will create a spinning vortex of rising hot air and gases that will carry smoke, flames, and debris thousands of feet into the air. These *fire whirls* can be hundreds of feet wide, have the intensity of a small tornado, and reach temperatures in excess of 2000°F.[16]

Some common firefighting terms that might be helpful for the animal first responder include the following:

1. Anchor point. This is a point where firefighters will start to construct a fire line. The anchor point is used to minimize the chance of being flanked by the fire while the line is being constructed.
2. Control line. Constructed or natural barriers and treated fire edges used to contain a fire.
3. Fire line. The part of a containment or control line that is scraped or dug to mineral soil.
4. Mop-up. Extinguishing or removing burning material near control lines.
5. Contained. The status of a wildfire suppression action signifying that a control line has been completed around the fire and any associated spot fires, which can reasonably be expected to stop the fire's spread.

[15] https://www.nps.gov/articles/wildfire-causes-and-evaluations.htm.
[16] https://www.livescience.com/45676-what-is-a-firenado.html.

6. Controlled. The completion of control line around a fire, any spot fires, and any interior islands to be saved. Firefighters may burn out unburned area adjacent to the fire side of the control lines to prevent additional spread.

The fire service will never knowingly put a first responder's life at risk, but the animal evacuation team may be working in some extremely dangerous conditions, and it's critical that they be aware of those situations that might put them at risk. As mentioned at the beginning of this chapter, the goal for any first responder program is to learn those skills that will keep the responder and their teammates safe. S-190 will cover two different lists that will hopefully keep the evacuation team out of harm's way: Standard Firefighting Orders and 18 Watchout Situations.

Standard Firefighting Orders:

1. Keep informed on fire weather conditions and forecasts.
2. Know what your fire is doing at all times.
3. Base all actions on current and expected behavior of the fire.
4. Identify escape routes and safety zones and make them known.
5. Post lookouts when there is possible danger.
6. Be alert. Keep calm. Think clearly. Act decisively.
7. Maintain prompt communications with your forces, your supervisor, and adjoining forces.
8. Give clear instructions and insure they are understood.
9. Maintain control of your forces at all times.
10. Fight fire aggressively, having provided for safety first.

18 Watchout Situations:

1. Fire not scouted and sized up.
2. In country not seen in daylight.
3. Safety zones and escape routes not identified.
4. Unfamiliar with weather and local factors influencing fire behavior.
5. Uninformed on strategy, tactics, and hazards.
6. Instructions and assignments not clear.
7. No communication link with crewmembers/supervisors.
8. Constructing line without safe anchor point.
9. Building fire line downhill with fire below.
10. Attempting frontal assault on fire.
11. Unburned fuel between you and the fire.
12. Cannot see main fire, not in contact with anyone who can.
13. On a hillside where rolling material can ignite fuel below.
14. Weather is getting hotter and drier.
15. Wind increases and/or changes direction.
16. Getting frequent spot fires across line.
17. Terrain and fuels make escape to safety zones difficult.
18. Taking a nap near the fire line.

A third list was developed to help firefighters remember the key safety factors and can easily be remembered using the acronym—LCES. Some departments have added an A for LACES:
"L"ookouts, "C"ommunications, "E"scape Routes, and "S"afety Zones.

Equipment

PPE: Animal first responders need to wear the same personal protective equipment that wildland firefighters wear. That will include a hard hat, face and neck shroud, eye and mouth protection, shirt and trousers, gloves, and boots. Fire PPE, even for the animal first responder, must meet the National Fire Protection Association (NFPA) 1977 standards.[17] Many evacuation teams will also require that their members carry a fire shelter. This is a small pup-like tent that the first responder can crawl into if overcome by fire.

---

[17] https://www.nfpa.org/codes-and-standards/all-codes-and-standards/list-of-codes-and-standards/detail?code=1977.

VOLCANIC ERUPTION RESPONSE

Deploying a team of animal search and rescue technicians in an area with active volcanic eruptions requires adequate PPE including respiratory protection. That may include disposable respirators for inside sheltering staff and cartridge respirators and full-body protection for the field teams.

In 2010, the eruption in Iceland under the ice cap, referred to as Eyjafjallajökull, disrupted air travel in northern Europe for several weeks and created a toxic threat to southern Iceland livestock. The concern for the horses and cattle was inhaling or ingesting the fluoride-laden ash. In high concentrations, fluoride becomes toxic as it turns to acid in the animals' stomachs, which can lead to hemorrhaging in the intestines. The Icelandic Ministry of Fisheries and Agriculture reached out to the International Fund for Animal Welfare to provide a subject matter expert on the best way to evacuate 400 Icelandic ponies from the areas impacted by the ash (Fig. 11.31).

FIG. 11.31 Icelandic horses outside the ash field, 2010. *Courtesy of The International Fund for Animal Welfare. All rights reserved.*

The SME conducted an assessment and met with the ministry to discuss options. In this case, there were three viable options: to round up and move the horses as a herd out to a safe zone, to trailer out the horses, and to shelter in place. Each of the options had advantages and disadvantages. The following photo shows the challenges of working within the ash field. Visibility was down to maybe 1 ft. The only way to see where the horses were inside the ash area was to turn on the flash of a camera. Seeing the red eyes of the horses—often just feet away—was the best way to locate them.

With low visibility and cross fencing throughout the area, the only option for herding the horses was to bring them down the main roads—something that the horses had never experienced—and there was some concern given that the horses were not shod. There were very few commercial trailers available, and when the group looked at the number of trucks and trailers that would be needed for that many horses, that option was determined not to be viable for that many horses. Interestingly, very few of the horses had ever been trailered, and given the stressful environment, it proved extremely challenging getting horses into a trailer.

Sheltering in place was an option for families with well-constructed barns where the horses could be closed in for a long period of time. Unfortunately, Iceland had experienced very mild winters for a number of years, so many of the horses had never been inside the barn and were not all that excited about going in. After a number of rousing discussions and debates, it was decided to provide all three options to the ranchers. The majority opted for relocating at a community center out of the ash field, and cowboys were able to herd a significant number down the highway without incident. It's seldom that the first plan works when dealing with animals as demonstrated in the Iceland response. During the Great Mississippi Flood in 2008, the author was working on option seven before coming up with one that finally worked!

It's important to carry a large toolbox and a good deal of patience and perseverance when working with animals in crisis.

EARTHQUAKE RESPONSE

Earthquake response typically involves technical rescue equipment and personnel to extricate victims from the rubble. The animal response team may be asked to provide expert animal handling or simply to be available to pass off any animals found for sheltering. Following the Haiti earthquake in 2010, animal response teams primarily provided recovery support to the nation in the forms of animal vaccinations and supporting the rebuilding of the veterinary infrastructure. And following the Sichuan earthquake in 2008, IFAW was requested to provide rabies vaccines to the large number of roaming dogs—a concession made to the government so that they would stop killing strays.

TORNADO RESPONSE

The level of destruction from a tornado like Joplin (2011) and Oklahoma City (1999) will be beyond anything that most responders could ever imagine. Entire structures are demolished, vehicles picked up and moved hundreds of feet away, and square blocks or even miles of dangerous debris scattered about putting rescuer and animals at risk. The immediate response to a tornado—like most events—will be to assess the level of damage and to determine the response objectives. Responder safety is number one, and in most cases, that will be to ensure that nontechnically trained responders stay out of the debris piles. In addition, it's critically important that all responders are wearing appropriate clothing and have the necessary equipment to keep themselves safe and to keep the animals in their charge safe. Typically, the animal response teams will be allowed into the impacted areas 24–48 h after the event depending on the level of destruction, safety concerns, and known animal concerns. Animals will survive even catastrophic tornadoes like what is shown in Fig. 11.32 and the Joplin Tornado. Scores of dogs and cats were rescued from the impacted area in the days following the tornado.

AHA was requested by the state of Georgia to assist in the recovery efforts in Camilla, Georgia, following a tornado outbreak that struck the area on February 13 and 14, 2000. When the team arrived to the ICP, they were asked by the state police why they were there. They explained that they had been tasked by the state to help coordinate animal rescue. The patrolman assured them that all of the animals had been removed by first responders. Fortunately, the team was allowed to set up their operations and over the next 5 days rescued 22 dogs (including a mom and 8 puppies), 12 cats, and 2 goats. The goats were found several hundred yards away from their home—in an attic space over a garage where the roof had been torn off.

Animals will survive, and you may find them in the strangest places.

On September 20, 2000, a series of tornadoes slammed into the town of Croton, Ohio. The tornado destroyed several warehouse buildings home to more than 1 million egg-laying hens. This operation used long rows of cages (literally packed with adult birds), placed in stepped-up tiers and suspended over an 8 ft manure pit to accommodate easy

FIG. 11.32 ASAR team following Joplin Tornado, 2011. *Courtesy of The American Society for the Prevention of Cruelty to Animals. All rights reserved.*

FIG. 11.33 Croton, OH tornado 2000. *Courtesy of American Humane. All rights reserved.*

removal of manure. This type of structure might enhance bird management and egg delivery, but it creates a death trap if a natural disaster occurs. When the buildings' sides and roofs blew down, the cages were mangled, and many birds were instantly killed; an equal number were trapped without access to food, water, or veterinary attention.

Various recue groups stepped up immediately to help and were able to take out a few hundred, but there were tens of thousands trapped in their cages and inaccessible without specialized equipment and a technical rescue team. There was simply no *quick* way to safely rescue them. Even if the birds could have been extricated quickly, large-scale in-state movement was difficult for a number of reasons and out-of-state movement difficult because of health concerns. It was one of those situations where the lead agency simply had to accept the fact that they were not going to be able to save all of the birds.

In that case, humane depopulation was the only humane recourse. An entire book could be written on these kinds of difficult situations that will come with animal rescue, but at the end of the day, the rule is to do what is best for the greatest number of animals. What made this situation even more difficult was trying to develop a procedure for bringing the cages up and away from the manure pit. And once the cages were pulled away, they were still so mangled that access to the inner rows was next to impossible. There was simply no way to reach the birds in a delicate, timely way (Fig. 11.33).

KEEPING AN EYE ON THE WEATHER

Rescue teams will often be working an area while dangerous weather is still around. It is always a good idea to have a portable weather monitoring device or to check in frequently with IC to determine what weather might be coming your way. There are a number of applications for smartphones that will provide radar images and alert you when severe weather is approaching.

Watching the skies and being aware of the types of conditions you might experience if severe thunderstorms are approaching may save lives. Be on the alert for dark, often greenish sky; large hail; large, dark, low-lying cloud (particularly if rotating); and a loud roar, similar to a freight train. If a tornado is approaching, consider the following:

- Find a safe room, basement, storm cellar, or lowest level.
- If no basement, go to the center of small interior room on the lowest level (closet or interior hallway) away from corners, windows, doors, and outside walls.
- Put as many walls as possible between you and the outside.
- Get under a sturdy table and use your arms to protect your head and neck.
- In a high-rise building, go to a small interior room or hallway on the lowest floor possible.

- Put on sturdy shoes.
- Do not open windows.

The ASPCA suggests that the animal first responder be aware of the following potential tornado hazards[18]:

- Hazardous driving conditions
- Slips and falls due to slippery walkways
- Falling and flying objects such as tree limbs and utility poles
- Sharp objects including nails and broken glass
- Electric hazards from downed power lines
- Falls from heights
- Exhaustion from working extended shifts
- Heat and dehydration

The animal rescuer working in tornado response or in any situation that requires an understanding of building structures should consider completing the following trainings:

- Compromised structure
- Confined space
- Trench rescue
- High- and low-angle rescue

In addition, the rescuer should use the following protective equipment:

- Helmet
- Gloves
- Boots
- Lighting
- Communications
- Eyes/nose/mouth protection
- Protective clothing

NUCLEAR ACCIDENT RESPONSE

The Three Mile Island reactor accident in 1979 had no detectable health effects on plants or animals, but it resulted in a number of changes involving emergency response planning. The 9/11 attacks provided further impetus for the federal government to develop standards that would ultimately reduce the risks of a nuclear accident.

In the United States, 100 commercial nuclear power reactors are licensed to operate at 62 sites in 31 states. Each site is required to have an on-site and off-site emergency plan. FEMA and the US Nuclear Regulatory Commission (NRC) are jointly responsible for the oversight of emergency preparedness for licensed facilities. The NRC has statutory responsibility for the radiological health and safety of the public by overseeing on-site preparedness and has overall authority for both on-site and off-site emergency preparedness.

Before a plant is licensed to operate, the NRC must have "reasonable assurance that adequate protective measures can and will be taken in the event of a radiological emergency."[19] Licensees and area response organizations must demonstrate they can effectively implement emergency plans and procedures through periodic evaluated exercises. Each plant operator is required to exercise its emergency plan with off-site authorities at least once every 2 years. Adding to the potential risk of a failure or accidentally release is the relative proximity of nuclear power plants with known fault lines (Fig. 11.34).

Whether exposure comes from a nuclear rod meltdown or a terrorist attack, humans and animals will be significantly impacted. An improvised nuclear device (IND) is a type of nuclear weapon. When an IND explodes, it gives off four types of energy: a blast wave, intense light, heat, and radiation. The bomb dropped on Hiroshima, Japan, at the end of World War II is an example of an IND. An improvised nuclear device would likely cause more nuclear fallout because of incomplete fission.[20]

[18] https://www.aspcapro.org/webinar/20170421/respond-tornadoes.

[19] https://www.nrc.gov/reading-rm/doc-collections/fact-sheets/3mile-isle.html.

[20] https://emergency.cdc.gov/radiation/pdf/infographic_improvised_nuclear_device.pdf.

FIG. 11.34 States with nuclear power plants. *Courtesy of United States Nuclear Regulatory Commission.*

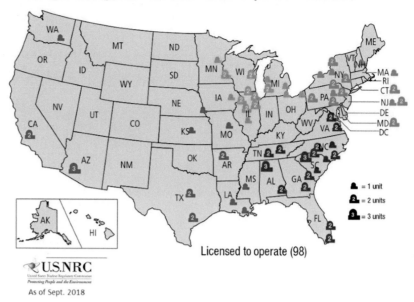

U.S. operating commercial nuclear power reactors

= 1 unit
= 2 units
= 3 units

Licensed to operate (98)

U.S.NRC
As of Sept. 2018

A number of regional and national committees have been established over the last 10 years to address animal issues following a nuclear emergency including but not limited to the following:

- Identifying health risks for animal responders
- Establishing safe evacuation zones
- Animal monitoring protocols
- Decontamination procedures
- Medical treatment interventions
- Protecting food supply
- Transport requirements
- Emergency sheltering needs
- Reunification process

Animal response teams are typically well versed in the rescue, sheltering, and reunification processes under *normal* conditions, but there is a paucity of information and trainings available to the animal rescue teams regarding potential radiation exposure.

Health Risks for Responders

Animal responders will not likely be asked to enter a hot zone but will be active in the monitoring and decontamination processes as animals are brought out of the evacuation zone. This will require PPE appropriate for the situation (zone) and constant monitoring of the radiation through the use of detection equipment. Detection equipment does not offer any protection against radiation; rather, it detects and tracks the levels of exposure. There are many types of radiation detection devices, but no single device can detect all kinds of radiation, and no one device is useful in all situations. There are four main types of radiation detection devices used by first responders:

- *Personal radiation detectors* is a small pager-like device that will quickly alert the wearer of the presence of radiation.
- *Dosimeters* measure the *accumulated* dose that the first responder has received while wearing the device.
- *Identifiers* determine the type of radioisotope that is present.
- *Survey meters* measure radiation levels of specific types of radiation.

The International Fund for Animal Welfare opted to use a badge-type dosimeter when they responded to the Japan earthquake and resultant nuclear accident (Fig. 11.35).

The government of Japan began distributing potassium iodide 4 days after the release of radioactive material.[21] Potassium iodide (KI) is a salt of stable (not radioactive) iodine that can help block radioactive iodine from being

[21] https://www.ncbi.nlm.nih.gov/books/NBK253930/.

FIG. 11.35   Dosimeter used by IFAW in Japan response, 2011. *Courtesy of The International Fund for Animal Welfare. All rights reserved.*

absorbed by the thyroid gland (part of the body most sensitive to radioactive iodine), thus protecting this gland from radiation injury. KI does not keep radioactive iodine from entering the body and cannot reverse the health effects caused by radioactive iodine once the thyroid is damaged.[22]

Establishing Safe Evacuation Zones

Following the Fukushima accident, evacuation orders were issued, and over time, the evacuation zones were expanded as more information became available. A number of families were evacuated out of one community only to get reevacuated after they had settled into another community. Initially, the evacuation zone was established at 2 km from the plant that was soon moved to 3 with shelter-in-place orders for anyone in the 3–10 km radius. All of these initial evacuation zones were set preemptively as there was no evidence that radioactive material had been released. Once there was evidence that material would be released, the evacuation zone was set at 10 km, and when Unit 1 exploded, the zone was once again reset to 20 km. An estimated 78,000 people evacuated from the 20 km "restricted zone."[23] By late April, the government no longer used a distance to determine an evacuation zone and rather used actually monitoring of radiation levels to determine restricted zones.

Animal rescue teams were allowed to work outside the hot (restricted) zones—in some cases assisting in the evacuation of families with pets and in other cases receiving animals from neighbors or friends and taking them back to temporary shelters. As the zones were pushed back, so was the work of the recognized (legitimate) rescue groups.

Animal Monitoring Protocols

As animals (and humans) exit the evacuation zones, they will be scanned to determine if they had been exposed to radiation. Radiation doses are measured in millisieverts (mSv). Fig. 11.36 shows that a dental X-ray exposes a patient to about 0.01 mSv and a full-body CT scan results in about a 10 mSv dose of radiation. Peak radiation level measured 4 days after Fukushima was 400 mSv/h, and for those Chernobyl workers who died within a month of the explosion, it was estimated that they were exposed to 6000 mSv of radiation.

Decontamination Procedures

There have been multiple attempts at designing rapid decontamination protocols for large populations of people and animals that have been exposed to radioactive material. Some of the challenges encountered include having adequate water supply, keeping temperature optimum for the fresh water, type of cleansing solution, disposal of contaminated water, amount of time per human and animal, handling fractious animals, and scanning protocols. The bottom line is that it is tough to come up with a system whereby hundreds of humans and animals can get thoroughly deconned in a reasonable amount of time. One can only imagine how frustrated folks would be if they were standing in Florida heat in August waiting for 12 h to get bathed.

[22] https://emergency.cdc.gov/radiation/ki.asp.

[23] https://www.ncbi.nlm.nih.gov/books/NBK253930/.

RADIATION DOSES AND EFFECTS

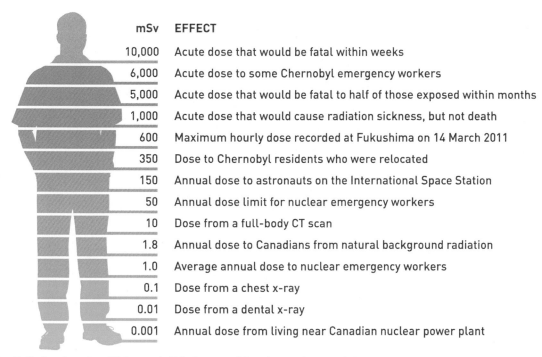

mSv	EFFECT
10,000	Acute dose that would be fatal within weeks
6,000	Acute dose to some Chernobyl emergency workers
5,000	Acute dose that would be fatal to half of those exposed within months
1,000	Acute dose that would cause radiation sickness, but not death
600	Maximum hourly dose recorded at Fukushima on 14 March 2011
350	Dose to Chernobyl residents who were relocated
150	Annual dose to astronauts on the International Space Station
50	Annual dose limit for nuclear emergency workers
10	Dose from a full-body CT scan
1.8	Annual dose to Canadians from natural background radiation
1.0	Average annual dose to nuclear emergency workers
0.1	Dose from a chest x-ray
0.01	Dose from a dental x-ray
0.001	Annual dose from living near Canadian nuclear power plant

FIG. 11.36   Radiation doses in millisieverts (mSv). *Courtesy of Canadian Nuclear Association.*

For small numbers of animals, the process is fairly straightforward. The animal is scanned; if they are below an acceptable level as set forth by local authorities, they can be moved directly to sheltering. If they exceed minimum levels, they go through the decon process. That process includes a complete bathing, rinse, and dry. The animal is scanned again. If they are below the minimum level, they go to sheltering; if not, they could go through a decon process a second time.

Medical Treatment Interventions

Animals suspected of being exposed but do not exhibit any signs of injury or illness are scanned and deconned before being triaged. If they have conditions that require immediate attention, they may be treated after scanning and prior to being deconned if the medical unit is set up to handle "hot" animals. As soon as the animal is stabilized, they go through the decontamination process.

Protecting Food Supply

One of the lessons learned from the Fukushima event was that as soon as livestock were provided untainted feed, their radioactive cesium levels decreased. Consequently, the primary management step post exposure is feeding animals uncontaminated feed and water.[24]

Emergency Sheltering Needs

Following the Fukushima event, the consortium of experts brought together by IFAW to develop best practices proposed that two emergency shelters be established—one outside the hot zone and one in the cold zone. The warm zone shelter would be an emergency shelter site for those animals that exceeded minimum exposure levels after two washes.

[24] https://link.springer.com/chapter/10.1007/978-4-431-55828-6_7.

FIG. 11.37 Dogs in colocated shelter Fukushima Prefecture. March 2011. *Courtesy of The International Fund for Animal Welfare. All rights reserved.*

FIG. 11.38 Pets sheltered in cars. Fukushima Prefecture. March 2011. *Courtesy of The International Fund for Animal Welfare. All rights reserved.*

The proposal was for these animals to stay in the shelter for 48 h and then be retested, and that process would continue until they reached minimum standards.

The cold shelter would be established to hold animals in the long term with the owners caring for their animals (Fig. 11.37). This shelter could also be used for colocating animals whose families were staying at the shelter. Colocated sheltering was not as well recognized in Japan as it is in the United States, but some of the larger human shelters were establishing pet shelters in the same facility so that owners would not have to keep them in their car (Fig. 11.38).

TECHNICAL RESCUE

In recent years, there has been a push to get animal rescuers more involved with technical rescue teams (TRT). Having someone who has experience in handling dangerous animals is a definite advantage when handling a fractious dog at the bottom of a dry well or managing a moose stuck in the ice. Technical rescue generally refers to situations where equipment and skills needed to execute a rescue fall beyond those normally associated with first responders. These situations might include rope rescue, structural collapse, confined space, ice, trench, and vehicle extrication rescues. Many larger fire departments will have a specialized tech team capable of performing complex rescues involving highly technical equipment, but smaller communities may be dependent on mutual aid.

Rope rescue is used in a variety of rescue situations. In essence, the higher the angle, the more technical the rescue. As the angle between the rescue team and the animal increases, the more difficult and dangerous it becomes to raise and secure them. There are three categories of rope rescue based on angle: low, steep, and high.

Low-angle rescue involves a rescue where the angle between the animal and rescue team is between 15° and 35°. In these situations, the weight of the animal is supported by the ground, and if the coefficient of friction between the victim and the ground can be reduced through a glide or board, then the amount of effort needed to raise is slightly more than the weight of the animal.

Steep-angle rescue involves situations where the angle between rescuer and victim is between 35° and 60°. In this situation, there is approximately as much weigh being distributed to the ground as to the ropes, and a greater mechanical advantage will be needed—or systems incorporated that will make it easier for the haul team to lift the animal.

High-angle rescue involves slopes that exceed 60° and may include a sheer drop between rescuer and victim. Here, the rescuers are completely dependent on rope systems to haul the victim to safety. This can be a risky situation—a misstep in the packaging of the animal or a faulty rope system could be fatal.

A compromised structure is one where there is not enough information available to determine whether the structure is safe for a first responder to enter. For the recent mudslides in Santa Barbara County, numerous structures were lifted off their foundations and carried hundreds of yards toward the ocean. At first glance, some of the structures looked intact, but those structures remained off-limits to first responders until an engineering team could evaluate the structural integrity. When a structure collapses with victims inside, rescuers are faced with a daunting task of removing rubble, shoring up access routes, and attempting to locate and rescue patients potentially buried under tons of rubble. Not only this is obviously a life-threatening situation for victims, but also it is a very dangerous situation for rescuers as they attempt to navigate a way through the debris.

Confined space rescue refers to a victim trapped in a narrow space or when getting to the victim requires gaining access through a narrow space. Underground vaults, silos, storage tanks, and sewers are examples of when a confined space rescue team might be called. This is a technically challenging situation as air quality is typically bad, there is poor lighting, communication is tough, and hazardous materials may be present. And like many situations involving technical rescue, time is of the utmost importance as available oxygen in confined spaces is limited. Unfortunately, that fact often leads to rescue attempts that are not well thought out and conducted by trained personnel. Two-thirds of all deaths occurring in confined spaces are attributed to persons attempting to rescue others.[25]

Trench rescue is a subset of confined space rescue and involves a situation where the depth of a trench is greater than the width and the width does not exceed 15 ft. Victims of trench collapse typically experience traumatic asphyxiation, and the impact of the collapsing material may cause broken limbs and internal injuries, so once again, time is of the essence. The potential for additional collapse is considered a primary hazard to rescuers. Removing soil or debris, adding weight near the edge of an open cut, vibration (such as vehicle movement), rain, or simply the passage of time may cause additional collapse at any time during the rescue operation.[26]

FIG. 11.39 Horse manikin on glide. *Courtesy of Toni MacPherson.*

photo: Toni MacPherson

[25] "Is it safe to enter a confined space?" (PDF). Cal-OSHA. 1998. Retrieved September 1, 2006.

[26] https://www.phoenix.gov/firesite/Documents/074768.pdf.

Ice rescue is a common occurrence in the United States in winter. Animals attempt to get out on the ice before adequate freezing or stay out on the ice too late in the season as melting is occurring. Like many other forms of technical rescue, ice rescue requires very specific personal protective equipment. Often referred to as gumby suits, the ice rescue technician must be properly fitted with a dry suit that protects them from the frigid water environment. Access to the victim and bringing a victim in over the ice requires specialized equipment to disperse the weight of rescuer and victim over a larger surface area. Time is once again against the victim, and the rescuer as an individual exposed to this environment may only have precious minutes to survive and keep their bodies above the ice.

We have seen a significant interest in large animal rescue over the last 20 years. The concept of raising a downed horse from a steep ravine is not new, but incorporating large animal rescue into fire-based technical rescue is a fairly recent advent. There are scores of large animal rescue (LAR) teams throughout the country, and many of those are community-based rather than fire-based. A number of Community Animal Response Teams (CART) and fire departments have taken it upon themselves to purchase the needed equipment and complete requisite training to be in a position to assist large animals in distress. In most cases, large animal refers to a horse, but the skills taught in LAR are applicable to many species.

There are a number of LAR courses available, but common topics include trailer accidents; technical large animal rescue from mud, water, ice, and low/high angle; and wildfire evacuations (Fig. 11.39).

12

Emergency Animal Sheltering

Emergency animal sheltering should only be undertaken when you have exhausted all other options. Following Hurricane Katrina, thousands of animals were "rescued." Many of those rescued animals were taken from backyards or even houses without significant damage. And there were even examples of where rescue groups took an animal from the backyard *after* the family had returned home. In many of those cases, it would have been better to provide food and water for the animal and allow them to stay at or near their home until their owner returned. Feeding-in-place programs require careful monitoring of the animal and the situation, but in most cases, the animal is happier, and the return-to-owner rate is much higher than with traditional emergency sheltering.

Shortly after Katrina, Hurricane Rita came roaring through SW Louisiana. There were a number of lessons learned from Katrina that influenced the response to Rita including controlling the number of groups responding. LSART was in the role of coordinating resources and assigned a single NGO to assist Calcasieu Parish and The barricades went up after they arrived to stop rogue groups from self-deploying. The other lesson learned from Katrina as referenced above was avoiding bringing in animals that could cope on their own in their existing location. Unlike Katrina, the real impact on pets following Hurricane Rita was the number of families that evacuated without them. The majority of flooding occurred south of the city, and the larger residential areas were relatively dry. Calcasieu Parish and the NGO (American Humane) developed a feeding-in-place program and only those animals that needed medical attention were brought into the shelter—and it worked beautifully. Utilizing large donations of food and the assistance from public works, Calcasieu Parish Animal Services kept their shelters open and the return-to-owner rate was nearly 100%.

There have been countless examples over the last 12 years where a feeding-in-place program was instituted and the advantages are numerous including:

- The shelters remains available for at-risk animals;
- That in turn reduces the number of staff needed for daily care;
- Animals that stay in their homes or in their territory are less stressed;
- Reunification rates are much higher—when the owner returns, his animal is already there; and
- One responding unit can generally complete 20–30 callbacks each day.

There are some disadvantages and times when a feeding-in-place program won't work:

- The home has experienced significant damage or the environment (such as mold) places the animal at risk or allows for escape;
- The owner cannot be identified or located;
- The hazardous conditions will persist for some time;
- The feeding areas are a great distance from the base of operations; and
- Large quantities of food are not available.

Feeding stations are much preferred for feral animals. Trapping a feral cat to bring it into a shelter and hold indefinitely is simply inhumane if they are capable of sustaining themselves in their natural habitat. Following the Sonoma fires (2017), only owned cats were trapped. If a known feral cat was accidently trapped, it was held for 3 days, sterilized, tipped, and returned to a safe (and familiar) area.

SHELTERING OPTIONS

When there is a need for sheltering animals, there are several options —each come with advantages and disadvantages. Conditions, agency policy, experience, resources, or timing typically drive the decision as to what type of shelter is used. Regardless of the type of shelter utilized, the primary goal is to provide quality daily care until the animal is reunited with their family or rehomed to a new family. There will be situations where animals rescued from the field will not be reunited, and a temporary (foster) or permanent new home will need to be found to avoid long-term sheltering.

After Hurricane Sandy came barreling through New York City, numerous shelters popped up throughout the region, and within two weeks, there were hundreds of unclaimed animals. New York City Emergency Management has done a spectacular job of embracing the animal welfare groups, and together, they have developed an Animal Planning Task Force (APTF) along with the Mayor's Alliance for NYC's Animals to ensure collaboration among the groups. HSUS, a member of the APTF, developed a foster program whereby people who needed to find a temporary home for the pets could look through a list of potential foster families. They would then work directly with that family to arrange a foster setting until they could get back on their feet. This was a brilliant idea that kept animals out of shelters and greatly increased the likelihood of families being reunited.[1]

Most jurisdictional areas will have a minimum hold period for animals brought in as strays. In most communities, that's 3–5 days. During disasters, the gold standard is 30 days. The response team must make every effort possible to try and reunite the animal with their family. If the owner is not located, after the agreed-upon holding time, that animal becomes the property of the AHJ and is available for adoption. Without exception, no disaster animal should be euthanized simply because of space and similarly, no shelter animal should be put down to make room for a disaster animal. There are simply too many other options available and too many groups willing to assist in a disaster for that to happen. The families deserve additional time, and responders have an obligation to exhaust every possibility. This is a member of a family that has likely suffered greatly, and reuniting them with their pet will go a long way toward the recovery and healing process.

Unfortunately, there can be a mind-set with responders that these animals must not have been loved and cared for or they wouldn't have been left behind. It's a common practice among volunteers to project their view on quality of care across all populations and scenarios. If an animal comes in that appears to be too lean, then the owners must not have taken good care of that animal. The responder doesn't recognize that there might be a host of medical reasons why that animal is thin. And consequently, if and when the owner is found, there is a hesitancy to return the animal to them. There have been numerous examples of where responders have stolen animals from an emergency shelter rather than see them returned to their legal owners. Discussing the mind-set and emotional challenges that the responder faces is certainly worthy of additional discussion but is outside the scope of this text. Suffice it to say that disaster response takes a toll on all of us—especially if we are charged with caring for children and animals!

Historically, animal-only shelters have an abysmal record of reunification. Following Hurricane Katrina, the Louisiana SPCA estimated that over 15,000 animals were rescued and 8500 of those were housed at Lamar-Dixon in Gonzales, LA—an animal-only shelter.[2] Best estimates for the percentage of animals reunited with their families were less than a quarter. That wasn't the fault of the shelter however. That was primarily the fault of a well-intentioned nation of responders that converged on New Orleans and "rescued" animals with little regard of them being reunited with their families. Some of the rescue groups were too self-absorbed to worry about what happened with the animal after they rescued it—that was somebody else's problem—they needed to get back into the field and rescue more.

Animal-Only Shelter (AOS)

In Animal-Only shelters, the care of the animal falls completely to the sheltering team (Fig. 12.1). There are a host of reasons why this type of shelter might be needed including

- Abandoned animals;
- Unowned animals;
- Owners not able to be located or have perished;
- Owners relinquishing their pet; and
- Owners not able to take care of their pet.

---

[1] https://www.aspca.org/about-us/press-releases/nyc-animal-planning-task-force-manages-reunification-efforts-pets-lost

[2] https://www.la-spca.org/katrina

FIG. 12.1   Animal Only Shelter (AOS), 2016. *Courtesy of The American Society for the Prevention of Cruelty to Animals. All rights reserved.*

Interestingly, there are many communities and a number of national groups that still prefer this type of sheltering. The sheltering team doesn't have to deal with family; they can better control the environment, and quite frankly, they can provide the level of care they feel is most appropriate. The staffing ratio for an AOS is approximately 10–15:1 depending on disposition, experience, and type and size of kennel. Veterinary support will be needed either at the shelter, at a nearby facility, or on call.

Co-Located Shelter (CLS)

As the name would imply, this is a sheltering situation where the family and pets are either in the same building, different room, adjacent building or nearby facility. Responsibility for the care of the animal falls to the owner. The sheltering team can assist as needed. CLS works most effectively when the owners accept the responsibility for caring for their animals. If done properly, a CLS requires few daily care staff. The approximate staffing ratio for a well-managed CLS will run 50–100:1. That suggests that with a shelter population of 300, three-six individuals should be able to ensure quality of care and that hygiene levels are maintained, equipment and supplies are readily available, and operational protocols are being followed. Veterinary support is usually available on call and at the owner's request (Fig. 12.2).

FIG. 12.2   Co-Located Shelter. Hurricane Matthew, 2016. *Courtesy of The American Society for the Prevention of Cruelty to Animals. All rights reserved.*

Co-Habitated Shelter (CHS)

In this situation, the owner and their pets are housed in the same area. A section of floor space is assigned to a family and they can configure it however they see fit. There are rules for containing their animals but in essence, they stay together as a family unit. The sheltering team has very little responsibilities and in most cases, depending on the agency in charge, shelter staff will be on call. If vet support is needed and not provided by the AHJ, then it will be the responsibility of the owner to contact and transport the animal as needed. In most larger disasters, there is a veterinarian available provide care for field and sheltered animals (Fig. 12.3).

Louisiana experienced one of the worst floods in its history in 2016. Two feet of rain fell in 48 h eventually resulting in 13 deaths and 60,000 homes damaged or destroyed.[3] Very quickly, human and animal shelters started popping up. Louisiana is no stranger to disasters and they have the sheltering process down pat. Interestingly, all three of the sheltering types discussed here were used following the floods at a single facility. Lamar-Dixon (Fig. 12.4) quickly became the site for large animals and many of the livestock owners brought their camping trailers and parked right next to the stalls they were using and in many cases, had their pets with them in their trailers and their livestock within feet of them. That's classic co-habitation sheltering. One end of Barn 1 was occupied by the Ascension Parish Animal Shelter. Their shelter received extensive damage from the flood. They evacuated their animals to Lamar-Dixon and stood up an animal-only shelter. About 400 yards from the barns was an ARC-run human shelter, and a co-located shelter was located in the main arena. So, on the grounds of Lamar-Dixon were well over 1300 animals being cared for either by their owners or by responders. The folks that had their animals with them gladly welcomed support of food and vet care but for the most part, all they needed was a roof over their heads. The co-located shelter had between 10 and 30 responders caring for animals where the owner was not caring for their animal properly or when they were not able to care for their animal (working, meetings, etc.). And the AOS was a never-ending whirlwind of activity with volunteers coming and going at all hours of the day trying to keep up with 100+ parish-owned pets.

In Baton Rouge, a spontaneous CHS appeared at the Celtic Studio[4] as 2000 people arrived (Fig. 12.5). There was no time and little effort taken to separate families and pets. They just needed a dry place to stay. That became the first state-supported CHS in Louisiana history and it worked amazingly well. Everybody got along famously, and there were very few incidents that required outside animal support. Interestingly, when assessment teams traveled from Lamar-Dixon to the Celtic Center, they didn't want to come back—it was so quiet and peaceful at the CHS. The only

FIG. 12.3 CHS Sonoma Fires, October 2017. *Courtesy of The American Society for the Prevention of Cruelty to Animals. All rights reserved.*

---

[3] http://fortune.com/2016/08/25/top-ten-fema-funded-disasters/

[4] This is for the photo: https://www.google.com/search?q=celtic+center+la+floods&tbm=isch&source=iu&ictx=1&fir=6xrsrneLoTjQBM%253A%252CqKdQL7YStPz7ZM%252C_&usg=__twMmZBsUPjM11_BGKFiryonBncQ%3D&sa=X&ved=0ahUKEwifi76_9v7YAhVSz2MKHQ_gDcUQ9QEIXzAK#imgrc=6xrsrneLoTjQBM:

FIG. 12.4 Lamar-Dixon Arena, August 2016. *Courtesy of The American Society for the Prevention of Cruelty to Animals. All rights reserved.*

FIG. 12.5 Celtic Studio. LA Floods, 2016.

real problem came when the state needed to return the Celtic Center to the owners for an incoming show. The folks without pets simply moved to another shelter but the pet-owning families were sent to a CLS and they were not happy about the decision. Complaints found their way to the commissioner of the Department of Agriculture and even to the governor demanding that they be able to stay with their pets.

The bottom line is that co-habitated sheltering is the gold standard. For the 2016 Cascadia Rising National Level Exercise, it was estimated that at least 50,000 pets would need to be moved from the west side of Washington to the east side if a major earthquake was to hit the Pacific Northwest. The Washington State Department of Agriculture stressed upon the exercise players that traditional sheltering would not work for this many evacuees. The only model that made sense was CHS. The plan that was finally agreed upon was to set up tent camps along the I-90 corridor in state parks. Families with pets would be in one part of the park in the same tent. Walking and playing areas would be established, and the family unit would stay together.

BASICS OF EMERGENCY SHELTERING

There are a number of excellent sheltering manuals[5] available to communities that are working on developing a sheltering plan. In addition, many of the national animal welfare groups have helpful tips on emergency animal sheltering. Consequently, this chapter will discuss some of the unique challenges that come with an emergency AOS and provide best practices and lessons learned from emergency shelters that were activated in recent years.

SHELTER MANAGEMENT

The Animal Sheltering Team[6] (AST) is responsible for the oversight, setup, operations, communication, and demobilization of a temporary companion animal shelter. The team is intended to provide a safe and protected environment for displaced companion animal populations. The AST:

- Ensures proper identification, tracking, reunification, and reporting of animals;
- Coordinates with Incident Command all facets of the animal response and intersecting components of the human response;
- Maintains safety and sanitation of the facility and equipment; and
- Ensures appropriate security is provided.

A Type 4 animal-only shelter (AOS) team would be capable of caring for 100 animals and be composed of at least nine responders, whereas a Type 1 AOS team would be capable of caring for 500 animals and include a minimum of 43 staff including a shelter manager; team leaders; animal care and handling specialists; animal identification, tracking, and reunification specialists; and animal behavior specialists. All of the management team members will have completed basic ICS and NIMS trainings at a level commensurate with their type. For example, Type 1 and Type 2 animal shelter managers would be required to have ICS 100-300 and IS 700 and 800.

SHELTER LOCATION

As mentioned earlier, the ASPCA was requested to set up an emergency shelter in St. Croix (Fig. 12.6) following Hurricane Maria (Sep. 2017). Initially, the advance team was looking for a place to house 100 animals, with room to expand to 200 if needed (over three months, 500 animals came through the shelter.) The ASPCA is well versed on what it takes to set up an emergency shelter but in this situation, there was no power, no water, and very few facilities that were intact to house this many animals. The Virgin Islands Department of Agriculture cleared out one of their open-air structures and the team set up their shelter using generators, water trucks, and a whole bunch of blue tarps. Fortunately, the majority of the roof was intact, but during monsoonal rains, it was tough to keep animals and responders dry. Logistics developed a designated team for irrigation control as small ponds quickly formed on the dirt floor.

Fortunately, the St. Croix situation is not the norm for disaster work in the continental United States. Generally, warehouse space or existing facilities are available within a reasonable distance of the impacted area. Appendix G provides a checklist that a sheltering team can use to identify key mechanical and physical aspects of a facility to ensure that it is appropriate for an emergency animal shelter. Important areas to consider are whether there is potable water, electricity, and good air flow and whether the building is structurally sound. Security, connectivity, parking, and loading docks are a few more areas to consider. Fairgrounds are a common site for emergency animal shelters, and there are a number of advantages and disadvantages with that type of facility. Just make sure not to schedule a disaster during fair season. The primary advantages to fairgrounds are the amount of space (which allows for small and large animals and co-habitated sheltering) and the prebuilt structures for livestock and fowl. The primary disadvantage is timing and seasonal challenges with temperature and airflow. In many communities, the fairgrounds are managed by a board not by county government, which can cause some challenges for securing and keeping a facility for long periods of time.

---

[5] http://lsart.org/sites/site-1707/documents/LSART_MANUAL_JUNE_2010.pdf; and http://www.flsart.org/pdf/countyFiles/Leon2.pdf

[6] Note FEMA—NIMS AES

INTAKE

Intake is the gateway to the shelter where basic information will be gathered about the animal and a unique identifier assigned (Fig. 12.7). Some emergency sheltering teams may use a paper system, whereas some groups will have access to a web-based electronic program. Even a simple excel document will work—or you may opt for a cardboard box with file folders.

Typically, animals will be assigned a unique identifier that will allow you at any time to be able to find out everything you need to know such as where it came from, when it was picked up or delivered, breed, gender, colorings, medical history, and microchip number. Some rescue groups will be able to start the identification process from the field either wirelessly or by calling dispatch with the information. In that case, they will be able to mark the kennel with that ID number either using a tag or on a piece of duct tape. This provides more accountability and enhances the opportunity for reunification. Once an ID number is assigned, it stays with that animal the entire time it is in custody.

Identifiers may range from neck bands to paint, but in most cases for dogs and cats, an inexpensive adjustable collar will suffice. These usually come in boxes of 500 bands and run less than $50. There may be a sticky tab or some way to adjust the length. Some dogs and cats are able to sneak out of them and if a tail of the band is exposed, their neighbors

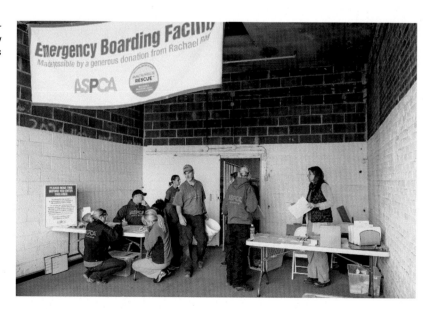

will eat it for lunch. In most cases, the unique ID number will be marked on the band with a permanent marker and it must be large enough (and legible) to be able to be seen from outside the kennel.

If animals are coming in from a large geographic area, the intake team may want to consider using an ID system that allows animals to be grouped by geographic area. For example, assume that there was a flood in Potter Valley and there were four distinct regions with the greater valley area. One possible numbering system might be PV-A-001. This type of system allows for expansion within a region rather than a straight sequential numbering system. Groups may also opt to create a system that includes an identifier by species. There are numerous ways to do it just remember rule number one—*keep it simple and manageable.* A generic intake form can be found in Appendix H.

TRIAGE

Triage is the process of sorting animals based on their immediate medical needs (Fig. 12.8). If they are asymptomatic and the vaccination history is known, the animal might be taken directly into the main shelter. If the shelter is unaware of the vaccination history and if space allows, the animal should be separated *by species* and *by age* from the main shelter to be observed. If the animal comes in exhibiting symptoms, the animal should be separated from the main population until deemed safe to enter. These are critical decisions to be made during the intake process. If an animal comes in that is critical, it may need to bypass the intake process and go directly to medical or isolation. In this case, intake will need to create the ID number and send a runner to medical to attach the band. The bottom line is that animals with obvious signs of illness or young puppies or kittens should not come into the general population. They need to be isolated, vaccinated, and cared for until they pose no threat to the rest of the shelter population.

Following Hurricane Sandy (2012), there were numerous emergency shelters set up throughout New York City and neighboring states. The ASPCA established an emergency shelter in the lower Bronx borough, and their numbers quickly swelled to over 300 animals. It became immediately apparent that a large percentage of animals were sick with upper respiratory infections and at one point, 80%–90% of the animals were symptomatic. The shelter team immediately established quarantine areas, initiated a very aggressive disease control plan, and fortunately did not lose any animals.

VACCINATIONS

Vaccinating animals at intake saves lives. Emergency shelters must develop a vaccination protocol that is tailored to their population's needs, and shelters must be ready to adapt their protocol if changes in overall

FIG. 12.8 Triage process Hurricane Sandy 2012. *Courtesy of The American Society for the Prevention of Cruelty to Animals. All rights reserved.*

population health are observed as part of a routine health monitoring program.[7] A number of groups (e.g., ASPCA, AAFP, and AAHA) have developed core vaccination guidelines for shelters. The resident veterinarian at the emergency shelter may modify those guidelines to fit the situation and available resources. In co-located and co-habitated shelters, it is recommended that owners provide evidence that their animal has been vaccinated before intake and if they are unable to do that, the pet should be vaccinated. There have been numerous examples during disasters where evacuees show up before the shelter staff does, and there may be scores or even hundreds of animals already in the shelter when the first responders arrive. In this case, every family must come through intake and either show vaccination records or be vaccinated. It may take a day or two to get everyone logged in and vaccinated, but if there is any chance that the shelter will be open for more than 48 h, then the delay and the chaos will pay off down the road.

Rabies vaccination on intake is not considered a priority in most shelters, as the risk of exposure is not high within most shelter environments. However, animals should be vaccinated against rabies when a long-term stay is anticipated, when risk of exposure is elevated, or when mandated by law. At minimum, animals should be vaccinated for rabies at or shortly following release.[8]

Primary Enclosure

A primary enclosure is an area to secure an animal such as a carrier, cage, run, kennel, stall, or pen. This is where an animal eats, sleeps, and in most sheltering situations spends the majority of its time. The two most common emergency shelter enclosures are a wire cage and a plastic crate (Fig. 12.9). Both are easy to transport and reasonably secure. The wire cage has a tray for easy floor cleaning and allows for better visibility and airflow. Given the wire structure, it is harder to completely disinfect. The plastic (airline) crates are easy to clean, better for transport, and stackable but have less visibility and airflow. For smaller dogs and puppies, you can double stack the wire cages. This allows for a change of scenery an opportunity to thoroughly clean the cage between switches, and some much-needed variety for enrichment.

Regardless of what enclosure you use, they must be secure, keep the animal safe and allow enough room for them to stretch, hold their tails erect, turn around, eat, and sleep. Some organizations utilize wire panels to form a 5′ × 5′ kennel with a gate (Fig. 12.10). These structures provide adequate space, ventilation, security, and visibility but are expensive and costly to ship. Cages and crates are adequate for short-term confinement, but larger spaces along with exercise areas will be needed if the response is expected to go beyond 7–10 days.

Daily Care

Emergency animal shelters can be a stressful place for both the animal and the caregiver. Animals may be stressed from the events preceding arrival, they are surrounded by scores of other (unknown) animals, it's loud and chaotic, and they are being handled by someone they have never seen before. The new caregiver may not have worked in this type of setting before, may not be an excellent handler, and may not use to being on their knees for 12 h. All in all, it's easy to see how this can be a stressful situation. Consequently, accidents happen. Dogs and cats strike out, animals escape, crates are not cleaned as well as they should be and animals get sick. But then, you get to the cage with the puppies and the kittens, and all is forgotten for a few precious moments.

Providing quality daily care is all about protocol and consistency. A protocol is a step-by-step process for getting something done. Protocols suggest a system of rules by which to work and thereby assuring a consistent process. Protocols for emergency animal shelters are written to meet recognized health standards to limit disease transmission and ensure quality of care. Protocols that were written for a brick and mortar building might not work with a situation where there is no hot running water or floor drains. Consequently, protocols may need to be tweaked to fit the situation. Emergency shelters that stay open for more than a few days will draw a large number of new people into the rotation. That rotation will likely include highly trained folks and some newbies. Responders that have been doing shelter work for some time may have a method that they like and that method may not be the same as the individual that worked that area the day before. The established protocol must be used every day regardless of the personnel.

---

[7] http://www.sheltervet.org/assets/docs/shelter-standards-oct2011-wforward.pdf

[8] http://www.sheltervet.org/assets/docs/shelter-standards-oct2011-wforward.pdf

FIG. 12.9   ASPCA shelter following Joplin Tornado, 2011. *Courtesy of The American Society for the Prevention of Cruelty to Animals. All rights reserved.*

FIG. 12.10   ASPCA AOS, Kansas City, MO, October 2014. *Courtesy of The American Society for the Prevention of Cruelty to Animals. All rights reserved.*

A protocol should be developed for every major facet of daily care and that protocol should be posted in clear view and be adhered to by all. The ASPCA has developed a protocol for every task in sheltering from washing bowls to cleaning cages (Fig. 12.11). As new responders rotate in, after the morning briefing, they go through a new responder orientation, assigned to a section, and that lead reviews all of the protocols associated with the tasks in that section. If a responder sees a situation where they think a protocol should be changed, they take that to the lead who takes it to the shelter manager for consideration.

The day in the life of emergency animal shelter daily care workers typically includes feeding, cleaning, exercising and daily enrichment—repeat as needed and times allows. Depending on the animal, type of enclosure, and support systems (e.g., running water), it may take 10–15′ to clean a kennel. Puppies can be slower, and cats can be quicker. Hopefully, there will be adequate staff to get the dogs out at least three times daily for walks, and a designated play area (preferably outside) will go a long way toward keeping the dogs from being bored and restless. A daily observation form has been provided in Appendix K.

FIG. 12.11 ASPCA AOS, St. Croix, USVI, 2017. *Courtesy of The American Society for the Prevention of Cruelty to Animals. All rights reserved.*

Behavior

When animals are going to be sheltered for longer than a few days, it's recommended that they have adequate opportunity to play, run, and enjoy some of the activities that they enjoyed when they were home. The role of the animal behaviorist is to determine what forms of enrichment will be best suited for each animal to lessen the stress associated with being kenneled for long periods of time. That may include daily romps in the exercise pen, kongs, frozen treats, toys, or more human and/or animal interaction (playgroups).

Animal behaviorists study the way animals behave and try to determine what causes certain types of behavior and what factors can prompt behavior change. They may have specialty areas such as fish, birds, large animals, wild animals, livestock, or household pets. They also may focus on certain types of behavior, such as hunting, mating, or raising offspring. Many things can influence how an animal behaves including long-term containment that can drastically alter an animal's behavior if steps are not taken to provide frequent enrichment activities.

The behaviorist is an important part of the intake and triage processes. They typically use a classification system to identify potentially dangerous animals or behaviors that could place the animal, handler, or other animals at risk. They are also a critical part of the reunification and rehoming teams as they provide valuable information on behavior and what type of home setting might be best for the animal and family.

In FEMA's Animal Resource Typing Guidelines, the *animal behavior specialist* assesses behavioral status, identifies potentially dangerous behaviors, and makes behavioral recommendations for animals during and post disaster. The animal behavior specialist has knowledge and experience working in one or more of the competency areas described below:

1. Companion animal including pets and service and assistance animals
2. Livestock including food or fiber animals and domesticated equine species
3. Wildlife/captive wildlife/zoo animal
4. Laboratory animal

A Type 1 animal behavior specialist

1. triages animals to determine behavioral status;
2. classifies animals by behavioral status to determine the appropriate type of animal care and handling specialist or other staff that can handle them;
3. identifies potentially dangerous behaviors that could put the animal, the handler, or other people or animals at risk;
4. designs and coordinates animal enrichment and socialization programs;
5. develops and implements treatment plans to address animal behavioral issues for stray, abandoned, unattended, or relinquished animals;
6. consults on appropriate postdisaster animal disposition.

Reunification

Emergency animal shelters should be a way station—a temporary facility to hold an animal until it goes home to its family. Unfortunately, an animal-only emergency shelter will likely end up with unclaimed animals. Experience has shown that not all pet owners are the same when it comes to the steps they will take to be reunited with their pets. For example, during the Los Alamos fire (2000), the individual that hiked 17 mi to reenter a hot zone to rescue his dog; the Black Forest Fire (2013) when a family drove their SUV through a burning fire to reach their lab; and, more recently, the Thomas fire (2017) when a maintenance worker ran into a burning barn to free horses. There are so many examples of truly heroic acts where people will do just about anything for their pets—they are one of the family. But all is not equal on that front. Some pet owners are irresponsible and do not properly care for their animals or purposely place them in situations where they are at risk and where the likelihood of reunification is low. And because of that, we will have unclaimed animals following an event.

This is such an important part of the response process that FEMA recently developed a job position for an animal identification, tracking, and reunification specialist that "provides intake, identification, tracking, and reunification support of animals with their owners or owner agents during various aspects of disaster response."[9] This individual will support all aspects of animal response including evacuation, sheltering, medical, and transport. The specialist will assist in the development of reunification protocols, the lost-and-found registry, and validation of animal reunifications.

The first step in the reunification process is bringing all of the active rescue groups together (Fig. 12.12). In major events, there will likely be numerous local, state, and national teams active in the response and sheltering efforts. The reunification process MUST be collaborative to be successful. Animals do not know geographic boundaries during a storm, and it's tough to know where an animal started its journey. The rescue groups need to share their information—preferably on a single website so that folks will not have to travel to a dozen shelters looking for their animal. Following the Butte Fire in Sonoma County (2017), a call was organized by all of the major players to coordinate efforts. Multiple jurisdictions and disciplines joined the nightly call—which continued for weeks after the event! One group was identified to collate and post all of the unclaimed animals on a single website with links to all of the other groups. It was an amazing success.

Social media is an excellent venue for reuniting animals with their owners. Posting on Facebook or other related sites have shown to be an excellent method for reuniting families. The team assigned to reunification will need to spend adequate time vetting the information that is on social media as it can be a blessing and it can be a nightmare wading (or weeding) through what is accurate and what is drama.

Establishing a hotline where folks can call when they do not have internet access has been found to be very effective as well. On numerous occasions for large events, the national groups have offered their hotline, which is staffed 24/7 to help communities organize the lost-and-found process. Even if power is up and running and internet is available, don't assume that the entire community will utilize that form of technology or even know where to start looking so it's important that the reunification process work off multiple platforms.

A microchip is oftentimes the ticket home for animals displaced during a disaster. It is a small device (size of a grain of rice) implanted between the animal's shoulder blades; a rather painless process similar to administering a vaccine.

FIG. 12.12   ASPCA Reunification Efforts following Joplin Tornado, 2011. *Courtesy of The American Society for the Prevention of Cruelty to Animals. All rights reserved.*

---

[9] Get actual site when FEMA approves

It is a radio-frequency identification (RFID) implant that provides a permanent identification number easily read by a scanner. It will last the lifetime of the pet. It's important to note that the owner information must be kept current and you need to register it if you want to be included on the various registries, which is a relatively quick and easy process.[10] If you have an animal that is microchipped, the following website will hunt the various registries to identify the owner: http://www.petmicrochiplookup.org/.

A relatively new tool to the reunification process is a face-recognition program called Finding Rover. This is a downloadable app that acts as a database for lost and found pets. A clear picture is taken, and once the image is uploaded, the app's facial recognition system tags the dog's unique features and stores them in the event the animal goes missing. *"Finding Rover sends out alerts through a variety of partners and with in-app messaging. Anyone who finds an animal that the app notes as missing takes a photo of the dog. That image gets sent to the database where it's compared with the facial recognition notes from the original owner. The owner's contact information pops up, allowing the finder of the animal to get in touch with the dog's owner."*[11]

The reunification team will be responsible for developing a protocol to ensure that the animal in their possession goes to the right home. It's difficult to imagine, but there will be people who will come to an emergency shelter to window shop. These are generally folks that are looking for a specific breed for a specific purpose. There are a number of steps that can be taken to ensure that the team has done due diligence to prevent that from happening.

Following the Moore Tornado in 2013, there were hundreds of displaced animals strewn throughout several jurisdictions and being cared for by numerous rescue groups. The City of Moore brought in a national group to assist sheltering at the fairgrounds. Oklahoma City Animal Welfare housed scores of animals at their shelter and utilized one of their partner agencies and another national group to help house unclaimed animals. In the middle of the two was another shelter with a local welfare group. And to compound the problem, typical of tornadoes, displaced animals crossed freely between the two jurisdictions. Shortly after the tornado, the OK Department of Agriculture Food and Forestry (DAFF) set up a triage center at the Incident Command Post and registered and transported animals to the various shelters. So, there were multiple jurisdictions dealing with hundreds of displaced animals.

All three of the major shelters started posting animals on their website, and a number of groups were using social media to post, but folks looking for their pet would need to visit at least three shelters to ensure that they had covered all of the sheltering options. DAFF stepped in to collate the various websites into a single site that folks could visit that helped immensely.[12]

The ASPCA was working with the Central Oklahoma Humane Society who was tasked by OKC Animal Welfare with sheltering and reuniting unclaimed pets. A trailer was moved to the front of the property—away from the shelter—to act as the reunification center (Fig. 12.13). A table was set up outside the trailer for owners to meet the team. Every owner was asked to show a photo of their pets—it's amazing how many carry a photo on their phone! If they did

FIG. 12.13   Reunification process following Moore Tornado, 2014. *Courtesy of The International Fund for Animal Welfare. All rights reserved.*

[10] https://www.preventivevet.com/dogs/how-to-update-your-pets-microchip

[11] http://www.findingrover.com/

[12] http://kfor.com/2013/06/07/registry-pets-lost-after-moore-tornado/

FIG. 12.14 Joplin Adoption Event, 2011. *Courtesy of The American Society for the Prevention of Cruelty to Animals. All rights reserved.*

not have a photo, they were asked to provide information about their vet. The vet was contacted to make sure that the individual was the rightful owner. After a complete description of the animal was provided, they were then allowed into the trailer to peruse photos of all of the animals in the shelter. If there was a connection or close resemblance, they were allowed in the shelter to see that animal. A member of the team escorted them through the kennels to monitor the reaction of the animal to the individual. Scores of animals were reunited and to the group's knowledge, there were no incidents where a pet went home to the wrong owner.

Unclaimed Animals

As noted previously, there may be unclaimed animals from an emergency sheltering situation. Teams should plan on a 5%–10% unclaimed rate whenever standing up an animal-only shelter or when performing animal search and rescue operations. As mentioned previously, the gold standard for holding disaster animals is 30 days and after that time or the time recognized by local ordinances, the animal will become the property of the AHJ. At that point, there are a number of options including fostering, adopting, or translocating the animal to another rescue group. The ASPCA has held a number of extremely successful adoption events following an emergency where hundreds of animals found new "forever" homes.

Following the devastating tornado in Joplin, MO (2011), the ASPCA held an adoption event that was attended by over 5700 people from 24 states and resulted in 745 unclaimed animals being adopted (Fig. 12.14). In 2016, the ASPCA assisted state authorities in North Carolina to seize over 700 animals. After exhausting all options for reunifying animals with the owners, the ASPCA held a 3-day adoption event that was so successful, that they ran out of adoptable animals by noon on day 2. ASPCA surveyed adopters 2–3 months after the event and found a 97% retention rate—a rate higher than found in more traditional adoption events![13]

PERSONAL PROTECTIVE EQUIPMENT

There will be situations in emergency animal rescue and sheltering where personal protective equipment (PPE) will be needed to protect the responder and the animals in their charge. PPE provides a protective layer between possible contaminates in the air or on surfaces and the clothes and skin of the responder. Much like requiring a dry suit for flood response, PPE should be required in emergency animal shelters whenever there is a concern for disease propagation or contact with a hazardous material.

Generally, the safety officer is responsible for determining the appropriate level of PPE, but with smaller response teams, the decision on the level of protective clothing might be made by the IC or the AHJ. Animal responders may not have experience donning and doffing PPE, so initial training and ongoing monitoring will be necessary to ensure the

[13] https://www.avma.org/News/JAVMANews/Pages/130701s.aspx

proper fit and understanding of the equipment's applications and limitations. Appendix L identifies the steps taken for safely putting on and taking off PPE in a high-risk environment.

The level of PPE required will be determined by the type and magnitude of the hazard. Floodwater response will require dry suits; working in and around the debris from a tornado will require mouth and eye protection along with boots, gloves, helmet, and protective clothing; and decontaminating a dog that has been exposed to radioactive waste will require full protective equipment. So each disaster response may bring about different levels of PPE requirements.

The National Institute for Occupational Safety and Health has identified a number of potential health and safety hazards associated with caring for displaced domestic animals. These include bites, scratches, and crushing injuries; exposure to zoonotic organisms and bodily fluids; injuries related to sharp, jagged debris; and heavy-lifting injuries. Their guidelines for PPE when handling animals include gloves, protective eyewear, durable clothing, and protective footwear.[14]

PPE and specifically Tyvek-type coveralls can be extremely warm and that will need to be taken into consideration when determining duty periods. Following the Deepwater Horizon oil spill in 2010, BP Oil contracted with Tri-State Bird Rescue and Research to oversee the washing of birds impacted by the oil. Tri-State requested the assistance of International Bird Rescue Research Center (now International Bird Rescue) and, with the assistance of a number of rescue groups including the Louisiana State Animal Rescue Team (LSART) and students from LSU Veterinary School, established rehabilitation centers in Louisiana, Alabama, Mississippi, and Florida. More than 8000 oiled birds were captured or collected (alive and dead), and 1246 were rehabilitated and released.[15] Even though it was a late spring event, temperatures in Louisiana were approaching high 80s, and with humidity levels in the 80% range, it felt more like 90°F. Add the protective equipment as seen in Fig. 12.15, and responders in the wash stations were restricted on how much time they could stay in a station before cycling out for required rest breaks.

There is a relatively lower risk for airborne contaminants than for skin contamination when handling animals, but it is advisable to require nose and mouth (respiratory) protection as well, and an N-95 particulate respirator is usually the best device. These respirators must form a tight seal around the nose and mouth to be effective, so each responder will need adequate guidance to ensure a proper fit.

Eye protection may include safety glasses, but when there is a concern for splashing (working around dangerous liquids), then tight-fitting goggles should be used. Full-face shields can be worn for major splashing, but safety glasses or goggles are still required.

On December 19, 2016, the New York City Department of Health and Mental Hygiene collected a respiratory specimen from a veterinarian experiencing influenza-like illness after exposure to sick domestic cats at an animal shelter in New York City. The virus was nearly identical (99.9%) to a virus isolated from a cat from a New York shelter where the veterinarian had worked. It was the only known report of direct transmission of influenza from a cat to a human, and needless to say, it triggered a major response from the public health community.[16]

FIG. 12.15 Pelican being washed from Deepwater Horizon oil spill, 2010. *Courtesy of International Bird Rescue.*

[14] https://www.cdc.gov/niosh/topics/emres/animals.html

[15] https://www.bird-rescue.org/about/history.aspx

[16] https://wwwnc.cdc.gov/eid/article/23/12/17-0798_article

FIG. 12.16 ASPCA Quarantine Facility in New York, 2016. *Courtesy of The American Society for the Prevention of Cruelty to Animals. All rights reserved.*

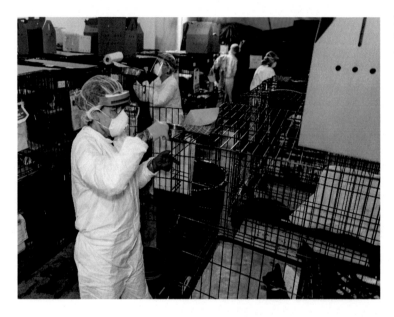

FIG. 12.17 ASPCA Quarantine Facility in New York, 2016. *Courtesy of The American Society for the Prevention of Cruelty to Animals. All rights reserved.*

Very soon after the detection of the virus, the ASPCA was requested by New York City Emergency Management and the NYC Animal Care and Control to establish an emergency quarantine facility to assist in controlling the outbreak of avian influenza (H7N2) in cats (Fig. 12.16). Establishing isolation and quarantine areas is nothing new to emergency sheltering groups but setting up a complete quarantine facility for what soon became over 500 cats over 3 months was new to the ASPCA. A warehouse was identified and the setup team came in to clean and establish cold, warm, and hot zones. *Cold* zones would be for responder rehab, logistics, and general movement of responders. The *Warm* zone was used for transitions into and out of the quarantine areas, and the *Hot* zone was the actual shelter.

On average, 100 responders were used each day to care for the cats and provide overhead management. Even though it was the middle of the winter, time in PPE was limited to 2 h, which meant that there were multiple changes of PPE daily. Even with bulk purchasing, each PPE unit (Tyvek coverall, gloves, face shield, booties, respirator, and hairnet) costs about $10, which equated to about $4000/day for PPE alone (Fig. 12.17).

DECONTAMINATION

The decontamination of animals and equipment exposed to hazardous materials is a critical piece of the overall animal emergency response. In most larger-disaster responses, Hazmat or Fire Services will stand up a decon station for vehicles and equipment but the decontamination of animals and personnel typically falls on the response team. Late in the Katrina response, one of the animal water rescue teams was supported by a local hazmat team, and they would meet the rescue team each afternoon and decon the dry suits, boats, and vehicles. If the team had *friendly* animals, they would decon them as well. Unfortunately, this was a relatively rare occurrence in the largest water-based animal rescue effort in US history. For those groups that were working within the command structure established for the primary shelter, *some* teams, after transferring their animals to intake, would drive their vehicles behind the shelter to decon their dry suits and equipment. This was hit-and-miss—at best. It's also important to note that the animal decon process for many of the intake centers was less than ideal with a lack of well-defined decon protocols and consistent procedures followed. A key lesson learned from the Katrina response was the need for more training in the decontamination process of animals and equipment.

The most often used decontamination agent for small animals and birds is liquid dish detergent, which is reasonably effective at removing many contaminants including oil-based products. As discussed in the washing process following Deepwater Horizon, the animal decontamination process is a difficult, back-straining, hot, and messy work. Consequently, it is not a team that is always easy to fill. It requires attention to detail and good animal handling experience.

Not all animals like taking a bath – especially from some stranger covered head to toe in white and looking like something out of a sci-fi movie.

It also requires significant attention to their own well-being, making sure that they are closely following personal protection protocols as they work (intimately) with animals potentially exposed to hazardous materials and zoonotic diseases. That weighs on responders over time; consequently, responders assigned to decon stations need to rotate regularly, and the safety officer needs to be vigilant on responder safety, including mental and emotion well-being and overall team morale.

Equipment Decontamination Protocol

Cleaning time is when responders are most likely to inadvertently spread disease to animals or people as they come into contact with hair, feces, litter, food, and maybe even a face washing from a quick tongue. Then, the responder has the opportunity to track whatever they came into contact with—throughout the shelter.

There is no one perfect product for sanitation in a shelter and a number of different products for cleaning and disinfecting different areas and objects may be needed. In most emergency shelters, where there is not a concern for exposure to hazardous materials or highly contagious diseases, deconning equipment and supplies typically entails removing all organic material, spraying on a germicide, allowing it to stay in contact with exposed surfaces for a designated period of time, and then thoroughly rinsing and drying. The ASPCA has an excellent webinar posted on their pro site, *Guidelines for Standards of Care in Animal Shelters*,[17] that provides the basic guidelines for ensuring a clean, safe sheltering environment. The chemicals involved with cleaning equipment have the potential to cause harm to both animals and people; therefore, written sanitation protocols must address animal and responder safety.

As previously mentioned, when dealing with known infectious diseases or when working in hazard sites, the emergency shelter will traditionally be divided into three zones—hot, warm, and cold. The hot zone is the area that contains the hazard, the warm zone is a buffer between the hot and cold zones (corridor), and the cold zone is the area free from any hazards. In NYC following the cat exposure to avian influenza, the ASPCA clearly marked these zones and placed a door between each area to reinforce the importance of not bringing contaminants into or out of a zone. The shelter was designed so that access into the warm zone was in an area that was easily monitored.

Two equipment decon stations were established for that operation (Fig. 12.18): one for equipment and supplies that would stay in the hot zone (water bowls, mats, toys, etc.) and another for equipment (cages and carriers) that would be removed from the shelter and taken back to the warehouse. The latter area was set up in a transition zone adjacent to the hot zone. Equipment was brought from the hot zone to the corridor, and after washing, it went into a warm area for rinsing and drying. It was then moved to the cold zone for repackaging.

The ASPCA, with assistance from a number of outside subject matter experts, developed an equipment cleaning protocol that has been included in Appendix M. The purpose of establishing a rigorous decon protocol is to ensure that no contaminant (or potential virus) is left on any cages or equipment before being repackaged and eventually reused.

[17] https://www.aspcapro.org/webinar/20110526/shelter-guidelines-sanitation

Animal Decontamination Protocol

The animal decontamination protocols used will depend on the nature of the disaster and the type of hazardous material exposure. A number of small animal decontamination protocols have been established, and for the most part, with minor modifications, these protocols can be used for large animals as well. In essence, the decontamination process involves leading contaminated animals through a series of sequential decontamination stations that each removes a particular type or category of contaminant.

The Animal Branch of the FEMA CBRNE Committee recommends that three decontamination stations be established before animals arrive. The first station is for removing any contaminated objects on the animal, such as leashes, collars, and halters. The second station is for washing the animal with detergent to remove organic matter, and the third station is for washing the animal with an antimicrobial solution to kill any microorganisms.

Animals and their owners should be photographed before entering the various stations and a unique identifier assigned. Leashes, collars, tags, and any other material that has been in contact with the animal should be photographed and removed at the first station. A new leash will be used to transfer the animal to the second station.

Station 2 will consist of two large basins with rigid shallow walls placed on a tarp on the ground. A spray bottle connected to a water hose providing fresh water along with various cleaning devices and liquid detergent will be available. Two responders will be needed at this station: one to handle and one to wash. The animal is washed in one basin and rinsed in the other.

The animal is washed thoroughly with a diluted detergent solution working the soap well into the hair. The animal is then rinsed and transferred to the third station. The washing and rinsing process will vary depending on the contaminant and the number of animals to decontaminate, but generally, 3–5 min should suffice. Prior to the next animal arriving, the wash and rinse water needs to be drained and the tanks cleaned.

An antimicrobial solution is applied in the final station using the same process as in station 2. Once the animal has been decontaminated and rinsed, it is transferred to a handler just outside the clean zone to be towel dried (Fig. 12.19). If decontamination was adequate, the animal can be permitted to enter the cold zone, but if there are residual contaminants, the animal goes back to the second station, and the process is repeated.

FIG. 12.19 Final stage of animal decon process. NASAAEP Katrina Commemorative Bootcamp, 2015. *Photo by Anne McCann, USDA, APHIS Animal Care.*

13

Climate Change and the Impact on Animals

INTRODUCTION

In December 2017, millions of people from around the world were stunned by a viral video[1] depicting a starving polar bear struggling on atrophied legs to make its way across the Canadian arctic, rummaging through refuse in search of something to eat (Fig. 13.1). Critics were quick to point out that climate change could not definitively be blamed for the dying bear's condition, but others noted that the risk of starvation from melting sea ice, driven by climatic warming, has a long history of being regarded as a threat to polar bears and other arctic species.[2] It is in great part the looming threat of sea ice loss that has led to polar bears being registered as a vulnerable species by the International Union for Conservation of Nature (IUCN) Red List. Despite uncertainty across predictive models, they have suggested that a >30% reduction in mean global polar bear population may be likely.[3]

However, population losses in subgroups of polar bears have already been recorded in excess of this figure. Polar bears living in the southern Beaufort Sea in Canada are thought to have undergone a population loss of around 40% between 2001 and 2010 and even with concerns over the accuracy of early population estimates, the decline seems unlikely to have been less than 25% and may have approached as much as 50%.[4] Exemplifying the harsh conditions that melting sea ice has inflicted on polar bears, the Beaufort Sea is also the location of the longest ever recorded polar bear swim. Less ice means more water, and in 2008, a female polar bear swam for 9 days straight, traveling across 426 mi of water and losing 22% of her body weight and her cub along the way.[5]

Certainly, this does not mean that all polar bears will meet this same fate. What it does mean is that the time for action for some of these bears has already passed. Paul Nicklen, one of the photographers who filmed the starving polar bear in the viral video, wrote, "This is what starvation looks like. The muscles atrophy. No energy. It's a slow, painful death."[6] The cruelty enacted on polar bears because of anthropogenic climate change is not singular to one species, but rather, these effects can be felt globally and indeed throughout the United States with potentially catastrophic impact. Climate change has already begun to effect nonhuman animals from all walks of life and classifications. What remains is appropriate identification and preparation for the rapidly evolving nature of climate change and the damaging impact felt by nonhuman animal populations as disasters and emerging threats develop and intensify.

---

[1] Nicklen, Paul [Paul Nicklen Photography]. (2017, December 5). [Video file]. Retrieved from https://www.facebook.com/paulnicklenphoto/videos/10155204590778364/.

[2] Gibbens, S. (2017, December 7). Heart-Wrenching Video Shows Starving Polar Bear on Iceless Land. *National Geographic*. Retrieved June 20, 2018, from https://news.nationalgeographic.com/2017/12/polar-bear-starving-arctic-sea-ice-melt-climate-change-spd/.

[3] Wiig, Ø., Amstrup, S., Atwood, T., Laidre, K., Lunn, N., Obbard, M., Regehr, E. & Thiemann, G. 2015. *Ursus maritimus*. The IUCN Red List of Threatened Species 2015: e.T22823A14871490. Retrieved June 20, 2018, from https://doi.org/10.2305/IUCN.UK.2015-4.RLTS.T22823A14871490.en.

[4] Bromaghin, J. F., Mcdonald, T. L., Stirling, I., Derocher, A. E., Richardson, E. S., Regehr, E. V., Amstrup, S. C. (2015). Polar bear population dynamics in the southern Beaufort Sea during a period of sea ice decline. *Ecological Applications*, 25(3), 634–651. doi:https://doi.org/10.1890/14-1129.1 (web archive link).

[5] Casselman, A. (2011, July 22). Longest Polar Bear Swim Recorded—426 Miles Straight. *National Geographic*. Retrieved June 20, 2018, from https://news.nationalgeographic.com/news/2011/07/110720-polar-bears-global-warming-sea-ice-science-environment.

[6] Nicklen, Paul [Paul Nicklen Photography]. (2017, December 5). [Video file]. Retrieved from https://www.facebook.com/paulnicklenphoto/videos/10155204590778364/.

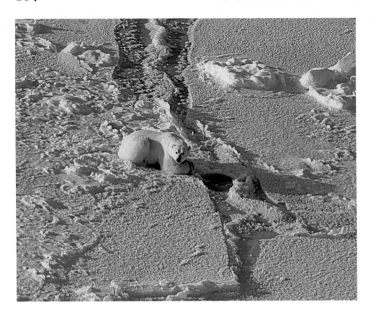

FIG. 13.1 An adult male polar bear still-hunting at a seal hole on the sea ice of the southern Beaufort Sea. *(Credit: Mike Lockhart, USGS. Public domain.)*

CLIMATE CHANGE AND IMPACT ON ANIMALS

Climate change has been identified as an emerging risk to global communities. According to the World Economic Forum's *The Global Risks Report 2018*, extreme weather events, natural disasters, and failure of climate change mitigation and adaptation have been identified as the first, second, and fifth top global risks, respectively, in terms of likelihood and second, third, and fourth, respectively, in terms of impact.[7]

The effects of global climate change will be felt in a number of different ways. According to the Intergovernmental Panel on Climate Change (IPCC), an increase of greenhouse gases in the atmosphere will likely increase temperatures over most land surfaces, though the exact change will vary regionally.[8] According to the IPCC, during the period covering 1880–2012, the globally averaged combined land and ocean surface temperature data show an approximate warming of 0.85°C.[9] NASA[10] puts this estimate a little higher, stating that the planet's surface temperature has risen about 1.1°C since the late 19th century. In addition to average temperatures, extreme temperatures will also be affected, with a greater probability of hot and record hot weather anticipated and an increase in temperature variability at both ends of the spectrum.[11] Anthropogenic climate change has also been projected to increase very large fire (VLF) potential in historically fire-prone regions in the United States. This will have greater impact in the intermountain West and Northern California through an increase in the frequency of conditions favorable to VLFs and an extension of the seasonal window.[12] Half of the twenty largest fires in California history have occurred in the decade between 2008 and 2018. In 2018, the Mendocino Complex fire surpassed 2017's Thomas fire to become the largest in state history by burning over 400,000 ac.[13]

[7] *The Global Risks Report 2018* (13th ed., Ref. No. 09012018). (2018). Retrieved July 11, 2018, from World Economic Forum's website https://www.weforum.org/reports/the-global-risks-report-2018.

[8] IPCC, 2014: Summary for policymakers. In: Climate Change 2014: Impacts, Adaptation, and Vulnerability. Part A: Global and Sectoral Aspects. Contribution of Working Group II to the Fifth Assessment Report of the Intergovernmental Panel on Climate Change [Field, C.B., et al (eds)]. Cambridge University Press, Cambridge, United Kingdom and New York, NY, USA, pp. 1–32.

[9] Hartmann, D.L., et al., 2013: Observations: Atmosphere and Surface. In: Climate Change 2013: The Physical Science Basis. Contribution of Working Group I to the Fifth Assessment Report of the Intergovernmental Panel on Climate Change [Stocker, T.F., et al. (ed)]. Cambridge University Press, Cambridge, United Kingdom and New York, NY, the United States.

[10] NASA. (2017, January 18). *NASA, NOAA Data Show 2016 Warmest Year on Record Globally* [Press release]. Retrieved May 27, 2018, from https://www.nasa.gov/press-release/nasa-noaa-data-show-2016-warmest-year-on-record-globally.

[11] NASA. (2005, March 30). The Impact of Climate Change on Natural Disasters. Retrieved May 12, 2018, from https://earthobservatory.nasa.gov/Features/RisingCost/rising_cost5.php.

[12] Barbero, R., Abatzoglou, J. T., Larkin, N. K., Kolden, C. A., & Stocks, B. (2015). Climate change presents increased potential for very large fires in the contiguous United States. *International Journal of Wildland Fire*. doi:https://doi.org/10.1071/wf15083 (web archive link).

[13] California Department of Forestry and Fire Protection. (2018). "The 20 Largest California Wildfires." Retrieved July 11, 2018, from http://www.fire.ca.gov/communications/downloads/fact_sheets/Top20_Acres.pdf.

As the global temperature rises, seawater will expand, and ice on land will melt, leading to a rise in sea level.[14] The elevation in sea level has led in large part to the marked increase of coastal flooding in the United States since the 1950s.[15] This trend is projected to continue and by 2050, the expected rise of 10–20 cm (4–8 in.) could more than double the flooding frequency in places such as the tropics.[16] Recent research has indicated that long-term sea level rise may have been underestimated, as it is now believed warming may be twice what prior models have suggested.[17] Warmer air is able to hold 7% more moisture per degree Celsius of warming, indicating that a warming world will also likely be a wetter one with increased precipitation.[18] Increases in global temperatures can result in increased intensity of storms, including tropical cyclones with higher wind speeds, a wetter Asian monsoon season, and more intense midlatitude storms.[19]

In general, rainfall related to tropical cyclones is proportional to rain rates and inversely proportional to the cyclone's translation speed. From 1949 to 2016, tropical-cyclone translation speed has decreased globally by about 10%, leading to increased rainfall totals.[20] So as more rain travels more slowly over a given area, higher rainfall totals should be anticipated. The damaging consequences of these sorts of developments have already been seen. In 2017, Hurricane Harvey made landfall in Texas, stalling partially over the waters of the Gulf of Mexico for 4 days and resulting in an unprecedented rainfall rather than moving inland and dispersing. Though Harvey is an extreme example, the impact of flooding can be felt throughout the United States. A review of FEMA's database of disaster declarations revealed that 73% of the presidential disaster declarations from 2008 to 2017 were flood-related events including hurricanes, severe storms, and heavy downpours.[21] During this same time period, eight of the ten states to have experienced the most flood-related presidential disaster declarations were inland, with Arkansas having the greatest frequency. Nevertheless, the potential for coastal flooding still remains significant. All of the top 10 states most at risk for devastating floods over the next 100 years are along the US coastline.[22]

Impact on Companion Animals

Few studies have been done on the direct impact of climate change on companion animals. Nevertheless, because of their close relationship with human populations, many of the effects of climate change felt by their human counterparts will be felt by the companion animals who cohabit and live among human populations worldwide.

Increased international attention has been given to the growing issue of climate refugees; those human populations displaced due to climate change and related disasters. The 2017 Internal Displacement Monitoring Center (IDMC)'s Global Report on Internal Displacement found that in 2016, there were 24.6 million new displacements from disasters. China had the highest number of displacement with over 7.4 million, the United States was fifth with over 1.1 million and small island states (like Puerto Rico) were found to "suffer disproportionately" after making considerations for population size.[23] As human populations are displaced, so are the animals who reside with them. Structures will need to be implemented within and between governmental and state entities in order to accommodate displaced individuals and their pets. An increased population of homeless pets should also be anticipated. Climate change affects social

[14] NOAA. (2009, June 5). How is sea level rise related to climate change? Retrieved May 12, 2018, from https://oceanservice.noaa.gov/facts/sealevelclimate.html.

[15] EPA. (2016, August). Climate Change Indicators in the United States: Coastal Flooding. Retrieved June 14, 2018, from https://www.epa.gov/sites/production/files/2016-08/documents/print_coastal-flooding-2016_0.pdf.

[16] Vitousek, S., Barnard, P. L., Fletcher, C. H., Frazer, N., Erikson, L., & Storlazzi, C. D. (2017). Doubling of coastal flooding frequency within decades due to sea-level rise. *Scientific Reports*, 7(1). doi:https://doi.org/10.1038/s41598-017-01362-7 (web archive link).

[17] Fischer, H., Meissner, K. J., Mix, A. C., Abram, N. J., Austermann, J., Brovkin, V., Zhou, L. (2018). Palaeoclimate constraints on the impact of 2°C anthropogenic warming and beyond. *Nature Geoscience*, 11, 474–485. doi:https://doi.org/10.1038/s41561-018-0146-0.

[18] Kossin, J. P. (2018). A global slowdown of tropical-cyclone translation speed. *Nature*, 558(7708), 104–107. doi:https://doi.org/10.1038/s41586-018-0158-3 (web archive link).

[19] NASA. (2005, March 30). The Impact of Climate Change on Natural Disasters. Retrieved May 12, 2018, from https://earthobservatory.nasa.gov/Features/RisingCost/rising_cost5.php.

[20] Kossin, J. P. (2018). A global slowdown of tropical-cyclone translation speed. *Nature*, 558(7708), 104–107. doi:https://doi.org/10.1038/s41586-018-0158-3 (web archive link).

[21] Lightbody, L., & Tompkins, F. (2018, January 25). Where It Rains, It Floods: Nationwide Disasters Underscore Need for Flood Policy Overhaul. Retrieved June 14, 2018, from http://www.pewtrusts.org/en/research-and-analysis/articles/2018/01/25/where-it-rains-it-floods.

[22] Koebler, J. (2012, March 14). 10 States Most at Risk of Flooding. *U.S. News & World Report*. Retrieved June 14, 2018, from https://www.usnews.com/news/slideshows/10-states-most-at-risk-of-flooding?slide=2.

[23] http://www.internal-displacement.org/global-report/grid2017/.

and environmental factors such as air quality, safety of drinking water, food sufficiency, and shelter security.[24] As individuals are displaced from their home communities, they may lose financial stability, food and/or water security, and the capacity to take care of themselves and their human and animal family members.

Short- and long-term health effects may also be felt by human and companion animal populations as a result of climate change. According to the World Health Organization (2018), between 2030 and 2050, an anticipated 250,000 additional human deaths per year may be attributed to climate change due to instances of malnutrition, malaria, diarrhea, and heat stress. US Global Change Research Program findings indicate that an increase of thousands to tens of thousands of premature heat-related deaths and a smaller decrease of premature cold-related deaths in most regions are projected each year as a result of climate change by the end of the century.[25] Despite the human tolerance for extreme temperatures having been observed to increase over periods of study ranging from years to decades,[25] at least one study done in the New York City metropolitan region suggests that acclimatization may not completely mitigate the effects of climate change, resulting in an overall net increase in heat-related premature mortality.[26] In addition to physiological acclimatization, increased use of air conditioning and other public health improvements, like cooling or heating centers, may help mitigate some of these impacts.[25] Despite these developments, there are still limits to either physiological or sociotechnical adaptations to heat.[25]

High ambient heat has been connected with illnesses affecting the cardiovascular, respiratory, and renal systems and has led to increased instances of illness and death from heat stroke, diabetes, hyperthermia, mental health issues, preterm births, cerebrovascular disease, and kidney disorders.[25] Since humans and nonhuman animals get many of the same diseases,[27] this would suggest that any negative health impact on humans may also be felt by companion animals.

Vector-borne diseases are also a national health concern that may be impacted by climate change. In the United States, there are currently 14 vector-borne diseases that are of national public health concern including West Nile virus, Lyme disease, and malaria.[28] Conditions altered by climate change are likely to affect waterborne and other types of diseases transmitted via cold-blooded animals, such as insects and snails, by lengthening the transmission seasons and altering their geographic ranges.[29] Disease outbreaks may also be impacted by climate change via altered "biological variables" including vector population size and density, vector survival rates, the profusion of diseased host animals, and the reproduction rates of the pathogens.[28] Uncertainty of the exact impact that climate alterations will have on vector-borne diseases may be due in part to the numerous factors that can interact and influence one another, including pathogen adaptation, changing ecosystems and land use, shifts in regional demographics, human behavior, and overall adaptive capacity.[28]

Lyme disease is exemplary of this development. While the disease is not tracked in companion animals in the way that it is tracked and reported in human populations, cases in companion animals should be expected to increase along with the recorded increase found in humans.[30] Fleas and ticks that carry the disease have also become smaller in size and more abundant, spreading Lyme disease into the now warming colder months and new counties and states across the United States.[30,31] Efforts stemming from continued mitigation techniques, such as vaccination or preventative flea and tick medication, may reduce overall impacts.

[24] World Health Organization. (2018, February 1). Climate change and health. Retrieved May 13, 2018, from http://www.who.int/en/news-room/fact-sheets/detail/climate-change-and-health.

[25] Sarofim, M.C., S. Saha, M.D. Hawkins, D.M. Mills, J. Hess, R. Horton, P. Kinney, J. Schwartz, and A. St. Juliana, 2016: Ch. 2: Temperature-Related Death and Illness. The Impacts of Climate Change on Human Health in the United States: A Scientific Assessment. U.S. Global Change Research Program, Washington, DC, 43–68. https://doi.org/10.7930/J0MG7MDX.

[26] Knowlton, K., Lynn, B., Goldberg, R. A., Rosenzweig, C., Hogrefe, C., Rosenthal, J. K., & Kinney, P. L. (2007). Projecting Heat-Related Mortality Impacts Under a Changing Climate in the New York City Region. American Journal of Public Health, 97(11), 2028–2034. doi:https://doi.org/10.2015/AJPH.2006.102947 (web archive link).

[27] Natterson-Horowitz, B., & Bowers, K. (2013). Zoobiquity: The Astonishing Connection between Human and Animal Health. New York, NY: Vintage Books, a division of Random House.

[28] Beard, C.B., R.J. Eisen, C.M. Barker, J.F. Garofalo, M. Hahn, M. Hayden, A.J. Monaghan, N.H. Ogden, and P.J. Schramm, 2016: Ch. 5: Vectorborne Diseases. The Impacts of Climate Change on Human Health in the United States: A Scientific Assessment. U.S. Global Change Research Program, Washington, DC, 129–156. https://doi.org/10.7930/J0765C7V.

[29] World Health Organization. (2018, February 1). Climate change and health. Retrieved May 13, 2018, from http://www.who.int/en/news-room/fact-sheets/detail/climate-change-and-health.

[30] Loyle, D. (2011, May 1). Spotlight on Lyme disease: Dr. Susan E. Little address epidemiology, treatment of this tick-borne disease. DVM360. Retrieved June 15, 2018, from http://veterinarynews.dvm360.com/spotlight-lyme-disease?id=&pageID=1&sk=&date=.

[31] Manning, S. (2016, February 15). More pests, heartworm for pets as world gets warmer. Express News. Retrieved June 15, 2018, from https://www.expressnews.com/news/bonus-content/article/More-pests-heartworm-for-pets-as-world-gets-6832126.php.

While these factors are primarily targeted toward and have been measured in terms of human impact, many companion animal populations living with or around human populations will likely feel these same effects. Climate-driven events that harm, kill, or displace humans will do the same to animals and potentially inhibit the ability of their human caretakers to care for them. Heat may also disproportionately affect companion animals, as many of the technological advancements to help mitigate climate change may not affect outside populations of companion animals like community and/or feral cats and dogs and companion animals kept outside for long durations.

Impacts on Livestock

More studies have been conducted to examine the impact of global warming on agricultural systems than companion animals. The global human population is expected to increase from 7.6 to 9.8 billion in 2050,[32] an increase of 29%, and demand for agricultural items to increase over 40%.[33] This population increase, when paired with increasing standards of living in developing countries, is expected to create a high demand for animal-derived protein by 2050.[34] With further intensification and growing demand, there is a need for further research into the impact of climate change on animal agriculture.

Temperatures are expected to increase across most land surfaces; extreme temperatures are also expected to increase, particularly extreme heat events. Thornton, Boone, and Ramirez-Villegas[35] summarized these and other direct and indirect impacts on production systems. Grazing systems may be directly impacted by extreme weather events, drought, floods, productivity losses due to temperature increase, and water availability. They further found that grazing systems may be indirectly impacted through issues related to fodder quantity and quality, disease epidemics, and host-pathogen interactions. In general, warmer conditions make disease transmission more likely between hosts, and new and emerging diseases can aid in both this transmissibility and the mixing of new genetic materials when they act as a "mixing vessel" between humans and livestock.[36] Nongrazing livestock production systems may be directly impacted by water availability and extreme weather events and indirectly impacted by increased resource pricing, disease epidemics, and an increased cost of animal housing due to the changing needs such as cooling systems.[35]

Rojas-Downing, Nejadhashemi, Harrigan and Woznicki[36] present the effects of climate change on livestock through three interlocking environmental shifts expected to emerge: increase of CO_2, increase of temperature, and variation in precipitation. Rather than focusing on grazing or nongrazing systems, Rojas-Downing et al. approach climate change impact on livestock from a climate-down perspective, recognizing the interlinkage between products of ongoing climate change. Specifically, regarding increasing temperatures, Rojas-Downing et al. raise concerns regarding water availability and increased resource consumption paired with decreased resource availability, reproduction, and health. Both an increase in temperature and an increase in CO_2 have the potential to impact the composition of pasture. Similarly, temperature and variation in precipitation may both increase the presence of pathogens and parasites and promote the spread of new and severe diseases.

The impact that climate change has on water availability and quality quickly emerges as one of the most significant issues for livestock globally. In the United States, agriculture accounts for approximately 80%–90% of consumptive water use.[37] Globally, the livestock sector uses one-third of the fresh water.[38] Increasing livestock populations to meet

[32] United Nations Department of Economic and Social Affairs. (2017, June 21). World population projected to reach 9.8 billion in 2050, and 11.2 billion in 2100. Retrieved July 11, 2018, from https://www.un.org/development/desa/en/news/population/world-population-prospects-2017.html.

[33] Alexandratos, N. & Bruinsma, J. (2012). *World agriculture towards 2030/2050: the 2012 revision*. ESA Working paper No. 12-03. Rome, FAO. Retrieved from: http://www.fao.org/docrep/016/ap106e/ap106e.pdf.

[34] Boland, M. J., Rae, A. N., Vereijken, J. M., Meuwissen, M. P., Fischer, A. R., Boekel, M. A., Hendriks, W. H. (2013). The future supply of animal-derived protein for human consumption. *Trends in Food Science & Technology*, 29(1), 62–73. doi:https://doi.org/10.1016/j.tifs.2012.07.002 (web archive link).

[35] Thornton, P. K., Boone, R. B., Ramirez-Villegas, J. (2015). Climate Change Impacts on Animals. (Working Paper No. 120). CGIAR Research Program on Climate Change, Agriculture and Food Security (CCAFS). Copenhagen, Denmark. Retrieved from https://cgspace.cgiar.org/bitstream/handle/10568/66474/CCAFSWP120.pdf?sequence=1.

[36] Rojas-Downing, M. M., Nejadhashemi, A. P., Harrigan, T., & Woznicki, S. A. (2017). Climate change and livestock: Impacts, adaptation, and mitigation. *Climate Risk Management*, 16, 145–163. https://doi.org/10.1016/j.crm.2017.02.001.

[37] USDA Economic Research Service. (2017, April 28). Irrigation & Water Use. Retrieved July 12, 2018, from https://web.archive.org/web/20171206185754/https://www.ers.usda.gov/topics/farm-practices-management/irrigation-water-use/background.aspx.

[38] Herrero, M., Havlik, P., Valin, H., Notenbaert, A., Rufino, M. C., Thornton, P. K., Obersteiner, M. (2013). Biomass use, production, feed efficiencies, and greenhouse gas emissions from global livestock systems. *Proceedings of the National Academy of Sciences*, 110(52), 20888–20893. doi:https://doi.org/10.1073/pnas.1308149110 (web archive link).

global demands exacerbates existing challenges of supplying water in many places, with some projections indicating a greater than 20% increase in cattle water demand by 2050.[39] The projected rise in sea level leading to potential increased saltwater introduction to freshwater aquifers could lead to water salination affecting animal metabolism, fertility, and digestion.[36] Water may also contain organic or inorganic chemical contaminants and heavy metals that have the potential to impair the cardiovascular, excretory, skeletal, nervous, and respiratory systems and impair production hygiene.[40] Any increased consumption by livestock due to increased temperature would likely exacerbate these negative health and production outcomes and overall water shortage.

The impact of heat stress on livestock is also significant, and the level of impact is dependent on temperature, humidity, species, genetic potential, life stage, and nutritional status.[36] Livestock in higher latitudes will be more affected by the increase in temperature than those in lower latitudes that may be better adapted to high temperatures and droughts.[36] Acclimation to heat stress can result in reduced feed intake and subsequent weight loss, increased water intake, and altered physiological functions such as reproductive and productive efficiency, change in respiration rate, alteration of glucose, protein and lipid metabolism, and changes in liver functionality.[40,36] In addition, heat-stressed animals may experience increased sensitivity to metabolic disease due to changes in organ function and subsequently experience negative outcomes for production, reproduction, and infectious disease sensitivities in intensive and extensive livestock production systems.[40]

Impacts on Wildlife

As exemplified by the polar bear, the effects of climate change on wildlife species have been broadly and increasingly researched in recent years. Efforts have not been focused solely on predicting or measuring potential future impact. A great deal of recent research has been focused on measuring the concrete impact that current warming has already had on various wildlife species. In general, climate change is responsible for four types of changes in species' traits: changes in density and range of the species, changes in phenology such as timing of migration or egg laying, changes in morphology like body size and behavior, and changes in genetic frequencies.[41]

Even with an average global warming of only around 1°C, scientists have recognized climate change impact across every ecosystem on Earth.[42] Out of 94 core ecological processes identified across terrestrial, marine, and freshwater ecosystems, 82% were found to have been impacted by climate change.[42] Negative impacts from climate change have also been measured for individual species, as many are threatened and may face extinction due to climate change effects. One study estimated that 47% of 73 species of terrestrial nonvolant threatened mammals and 23.4% of 1383 species of threatened birds have already been negatively impacted by climate change.[43] Climate change may also pose a threat to biodiversity, with hundreds of species already experiencing climate-related local extinctions, including 47% of 976 species studied by Wiens.[44] Though incidence was similar across climatic zones and habitats, extinctions were significantly higher in tropical species (55%) compared with temperate species (39%), animals (50%) compared with plants (39%), and in freshwater habitats (74%) relative to terrestrial (46%) and marine habitats (51%). Even small changes in species distribution can have a negative cascading effect on other species in a region.[44]

Climate change is impacting where animals reside. Most of the species that have shifted have moved to higher elevations or toward the poles, with animals requiring cooler temperatures shifting as their home temperature ranges rise.[45] One study estimated that species have recently shifted to higher elevations at a median rate of 11.0 m (36 ft)

[39] Porter, J.R., et al., 2014: Food security and food production systems. In: *Climate Change 2014: Impacts, Adaptation, and Vulnerability. Part A: Global and Sectoral Aspects. Contribution of Working Group II to the Fifth Assessment Report of the Intergovernmental Panel on Climate Change* [Field, C.B., et al., (eds.)]. Cambridge University Press, Cambridge, United Kingdom and New York, NY, USA, pp. 485–533.

[40] Nardone, A., Ronchi, B., Lacetera, N., Ranieri, M., & Bernabucci, U. (2010). Effects of climate changes on animal production and sustainability of livestock systems. *Livestock Science,130*(1–3), 57–69. doi:https://doi.org/10.1016/j.livsci.2010.02.011 (web archive link).

[41] Root, T. L., Price, J. T., Hall, K. R., Schneider, S. H., Rosenzweig, C., & Pounds, J. A. (2003). Fingerprints of global warming on wild animals and plants. *Nature, 421*(6918), 57–60. doi:https://doi.org/10.1038/nature01333 (web archive link).

[42] Scheffers, B. R., Meester, L. D., Bridge, T. C., Hoffmann, A. A., Pandolfi, J. M., Corlett, R. T., Watson, J. E. (2016). The broad footprint of climate change from genes to biomes to people. *Science, 354*(6313). doi:https://doi.org/10.1126/science.aaf7671 (web archive link).

[43] Pacifici, M., Visconti, P., Butchart, S. H., Watson, J. E., Cassola, F. M., & Rondinini, C. (2017). Species' traits influenced their response to recent climate change. *Nature Climate Change.* doi:https://doi.org/10.1038/NCLIMATE3223 (web archive link).

[44] Wiens, J. J. (2016). Climate-Related Local Extinctions Are Already Widespread among Plant and Animal Species. *PLOS Biology, 14*(12). doi:https://doi.org/10.1371/journal.pbio.2001104 (web archive link).

[45] IFAW. (2013). *Climate Change & Animals: How Climate Change Affects Life on Earth* (Rep.). Retrieved May 27, 2018, from https://s3.amazonaws.com/ifaw-pantheon/sites/default/files/legacy/education-publications/us/us-aae-climate-change-series-intro-life-on-earth.pdf.

per decade and to higher latitudes at a median rate of 16.9 km (10.5 mi) per decade, with the largest distances found in areas with the highest levels of warming.[46] This type of range shift was found in a selection of 35 nonmigratory European butterflies, where 63% had ranges that shifted northward by 35–240 km (22–149 mi) during this century and 3% that shifted southward.[47]

Though poleward and upward shifts are the most frequent types of range shifts reported, records exist of other types of range shifts being reported—such as east-west directions or even toward tropical latitudes and lower elevations.[48] Additionally, the rates of change for individual species can fluctuate greatly depending on multiple internal species characteristics and external forces of change such as rate of habitat loss.[46] These shifts have also been displayed to have far reaching, multifaceted repercussions, with evidence showing that regional and global species relocations due to climate change affect ecosystem functioning, human well-being, and the dynamics of climate change itself from increasing climate feedbacks initiated by alterations in albedo and CO_2 exchanges.[49] Increased flooding and very large fire (VLF) potential may also displace wildlife into geographic ranges or habitats outside of their natural range. This means they could be pushed into more human habituated areas and increase human and wildlife conflicts and distance wildlife from natural resources that they depend on for survival.

In addition to shifts in range, wildlife may experience shifts or alterations in certain behaviors as a result of climate change with potential deleterious effects. As with livestock, many animals may change their feeding and activity behaviors in an attempt to adapt to warmer temperatures. Elephants, for instance, eat less, rest more, and spend more time in the water and shade to cool down when temperatures rise.[50] Warming spring temperatures may cause bears to emerge from hibernation before their regular, primarily plant-based food sources are available, potentially causing them to starve or seek food in human populated areas leading to increased conflicts with humans. Birds may similarly begin their seasonal migrations or nesting earlier in spring.[45] One study of 65 Western European migratory bird species examined data collected over a 42-year period and showed that long-distance migrants wintering south of the Sahara have generally advanced their autumn passage while those wintering north of the Sahara have delayed their autumn passage.[51]

Diseases, disease vectors, and parasitic disorders that affect wildlife may change as a result of alterations primarily due to climate change. Infectious diseases impacting wildlife may see shifts as many pathogens are sensitive to temperature, rainfall, and humidity.[52] The French Agency for Food, Environmental and Occupational Health and Safety (ANSES) identified Rift Valley fever, West Nile virus infection, visceral leishmaniasis, leptospirosis, bluetongue, and African horse sickness as the six diseases likely to be the most affected by climate change.[53] In the United States, the Department of Agriculture APHIS Wildlife Services has explored the potential need for future involvement in managing wildlife diseases such as white-nose syndrome, aspergillosis, plague, Lyme disease, chronic wasting disease, Chagas disease, rabies, and hantavirus pulmonary syndrome due to potential climate-related changes.[54] And while some pathogens may decrease with climate change, most are expected to increase and may see changes in pathogen development and survival rates, disease transmission, and host susceptibility.[52] Increased frequency and proliferation

[46] Chen, I., Hill, J. K., Ohlemüller, R., Roy, D. B., & Thomas, C. D. (2011). Rapid Range Shifts of Species Associated with High Levels of Climate Warming. *Science, 333*(6045), 1024–1026. doi:https://doi.org/10.1126/science.1206432 (web archive link).

[47] Parmesan, C., Ryrholm, N., Stefanescu, C., Hill, J. K., Thomas, C. D., Descimon, H., Warren, M. (1999). Poleward shifts in geographical ranges of butterfly species associated with regional warming. *Nature, 399*(6736), 579–583. doi:https://doi.org/10.1038/21181 (web archive link).

[48] Lenoir, J., & Svenning, J. C. (2014). Climate-related range shifts—a global multidimensional synthesis and new research directions. *Ecography, 37,* 1–14. doi:https://doi.org/10.1111/ecog.00967 (web archive link).

[49] Pecl, G. T., Araújo, M. B., Bell, J., Blanchard, J., Bonebrake, T. C., Chen, I., Clark, T. D., Colwell, R. K., Danielsen, F., Evengard, B., Robinson, S., Williams, S. E. (2017). Biodiversity redistribution under climate change: Impacts on ecosystems and human well-being. *Science, 355*(6332), 1–9.

[50] WWF. (2018, March). *Wildlife in a Warming World - The effects of climate change on biodiversity in WWF's Priority Places* (Rep.). Retrieved June 1, 2018, from WWF—World Wildlife Fund website: https://www.wwf.org.uk/sites/default/files/2018-03/WWF_Wildlife_in_a_Warming_World.pdf.

[51] Jenni, L., & Kéry, M. (2003). Timing of autumn bird migration under climate change: Advances in long-distance migrants, delays in short-distance migrants. *Proc. R. Soc. Lond. B, 270,* 1467–1471. doi:https://doi.org/10.1098/rspb.2003.2394 (web archive link).

[52] Harvell, C. D., Mitchell, C. E., Ward, J. R., Altizer, S., Dobson, A. P., Ostfeld, R. S., & Samuel, M. D. (2002). Climate Warming and Disease Risks for Terrestrial and Marine Biota. *Science, 296*(5576), 2158–2162. doi:https://doi.org/10.1126/science.1063699 (web archive link).

[53] ANSES. (2016, August 25). The consequences of climate change on animal diseases. Retrieved July 4, 2018, from https://www.anses.fr/en/content/consequences-climate-change-animal-diseases.

[54] Algeo, T. P., et al., (2014). Predicted Wildlife Disease-Related Climate Change Impacts of Specific Concern to USDA APHIS Wildlife Services (R. M. Timm & J. M. O'Brien, Eds.). In Proc. 26th Vertebr. Pest Conf. (pp. 310–315). Davis, CA: University of California. Retrieved July 4, 2018, from https://www.aphis.usda.gov/wildlife_damage/nwrc/publications/14pubs/14-144 deliberto.pdf.

in vector-borne diseases and other zoonotic pathogens, including emerging infectious diseases—(EIDs), may result in population decline and further reduction in biodiversity (Daszak, Cunningham & Hyuatt, 2001; USGS, 2012).

PLANNING AND RESPONSE TO INCREASED CLIMATIC THREATS FOR ANIMALS

Emergency management has already, in many areas, accounted for animals in disasters, even those expected to increase in frequency, duration, severity, or overall impact due to climate change. However, without accounting for the rapid developments of these changes, appropriate mitigation, preparation, and subsequent response may lag behind. Agencies, organizations, and individuals must be prepared to rapidly respond to snowballing events that impact a multiplicity of animal groups; a wildfire that required wildlife rescue may lead to a mudslide requiring companion animal rescue. Climate change science and preparation will help targeted planning objectives in the emergency management field to better plan and subsequently respond and recover from anthropogenic-induced climatic events that impact humans and nonhuman animals alike.

The federal government framework for response in many ways remains unprepared to respond to the emerging and rapidly evolving threats and the breadth of animals with the potential to be impacted. Some initial steps have worked to advance the capabilities within the National Response Framework (NRF), such as the passing of the Post-Katrina Emergency Management Reform Act (PKEMRA) and the Pets Evacuation and Transportation Standards Act of 2006 (PETS Act). The PETS Act amends the Robert T. Stafford Disaster Relief and Emergency Assistance Act to ensure that state and local emergency preparedness operational plans take into account the needs of individuals with household pets and service animals prior to, during, and following a major disaster or emergency. Despite increased legislative and subsequent agency attention to nonhuman animals, risk-driven preparation linking nonhuman animals and climate change impact remains stagnated. As of this year (2018), FEMA has removed all reference to climate from their 2018–22 Strategic Plan (Gonzales, 2018; Lee, 2018; U.S. Federal Emergency Management Agency, 2018) detracting from the severity and focus that increasing climatic shifts will demand from the emergency management and response fields. The word climate does not appear at all in the document and emerging threats are attributed to terrorism and cybersecurity. It ignores one of the driving factors of change when it comes to the evolving nature of disasters presently and in the future.

Preparation at the state and county levels has primarily taken the form of animal response teams where individual animal welfare organizations, practitioners, and volunteers take center stage. The National Capabilities for Animal Response in Emergencies (NCARE) study published last year (Chapter 3) indicated that 31 of the 48 participating states (65%) reported having a State Animal Response Team (SART) (Spain, Green, Davis, Miller & Britt, 2017). The survey further revealed that 16 of 33 (48%) large counties with > 1 million population and only 131 of 565 (23%) reported having a County Animal Response Team (CART) with significant regional variation in the proportion of counties with teams ranging from 2% to 69% (Spain et al., 2017). Animal welfare organizations and practitioners have been at the forefront of the battle for amplified emergency animal preparedness in the face of growing impacts from climate change and should continue establishing a network of resources through collaboration with national animal response organizations and establishing mutual aid agreements (MAA) that take into account the increasing impacts of climate change and the potential for compounding disasters. Efforts should be focused on improving SART and CART saturation and enhanced through community emergency self-assessment supported by climate change assessments.

The increasing and evolving risk of climate change requires a paradigm shift from those responding to nonhuman animal-related emergencies in the face of disasters. As Adaptation Manager Missy Stults with ICLEI-Local Governments for Sustainability indicates, "for emergency planners and response personnel, it becomes really important to start planning for a changing paradigm. We can't plan based on historical situations anymore because history is literally being changed" (Pittman, 2010). If we wait for reactionary assessment and response planning, the potential for harm and loss of life remains substantial with the potential to increase as growing human and pet populations continue to move to the US coast, urbanize, and fall behind in infrastructure and large-scale mitigation efforts.

Appendices

APPENDIX A: FEMA ANIMAL RESOURCE TYPING

Resource Typing Definition for Situational Assessment
Animal Emergency Response

ANIMAL AND AGRICULTURE DAMAGE ASSESSMENT TEAM

DESCRIPTION	The Animal and Agriculture Damage Assessment Team performs initial and ongoing situational assessments and needs assessments to determine the types of emergency relief necessary in an incident involving animal and agriculture issues. This team's work begins in the early, critical stages of a disaster and continues throughout the incident. Depending on the incident and hazards, the team may assess community veterinary infrastructure, animal shelters, animal control facilities, feed stores, retail services, animal exhibitors (such as zoos), agricultural producers (farms, ranches, dairies, and so on), processors, market facilities, and agricultural support industries (agrichemical, feed mills, and so on). The team may also assess animal evacuation and shelter needs, stray animals, and animal carcass issues.
RESOURCE CATEGORY	Animal Emergency Response
RESOURCE KIND	Team
OVERALL FUNCTION	This team: 1. Photographs and records disaster site damage 2. Investigates locations where damage exists 3. Analyzes the significance of affected animal and agriculture infrastructures, crops, and animals 4. Estimates the extent of damages 5. Identifies potential cascading effects of animal and agricultural issues 6. Recommends initial priorities for response and recovery
COMPOSITION AND ORDERING SPECIFICATIONS	1. Discuss logistics for this team, such as security, lodging, transportation, and meals, prior to deployment. 2. The team typically works 12 hours per shift, is self-sustainable for 72 hours, and is deployable up to 14 days. 3. Requestor specifies specialty areas necessary, such as animal health professionals, agronomists, engineering specialists, logisticians, environmental experts, animal sheltering experts, communications specialists, and others

Each type of resource builds on the qualifications of the type below it. For example, Type 1 qualifications include the qualifications in Type 2, plus an increase in capability. Type 1 is the highest qualification level.

COMPONENT	SINGLE TYPE	NOTES
EQUIPMENT PER TEAM ASSESSMENT	1. Laptop computer(s) 2. Digital camera(s) 3. GPS capabilities 4. Office supplies 5. Measuring devices 6. Appropriate software 7. Cutting/trimming device(s) 8. Marking paint or other marking materials 9. Traffic control device (or safety signage) 10. Cleaning and disinfecting supplies 11. Other equipment and supplies as necessary, based on ordering specification	1. Appropriate software includes word processing, spreadsheet, Geographic Information Systems, and database management programs. 2. Measuring devices include tape measures (25' to 100') and measuring wheels. 3. Cleaning and disinfecting supplies are necessary when moving from one agricultural area to another. 4. Requestor provides Authority Having Jurisdiction (AHJ)-approved damage assessment forms for the team to complete.
EQUIPMENT PER TEAM COMMUNICATIONS	3 - Two-way portable radio 3 - Cell phone	Not Specified

Resource Typing Definition for Situational Assessment
Animal Emergency Response

COMPONENT	SINGLE TYPE	NOTES
EQUIPMENT PER TEAM MEMBER PERSONAL PROTECTIVE EQUIPMENT	PPE is mission specific and may vary by work environment; it includes: 1. Protective footwear 2. Protective clothing for skin exposure 3. Eye and ear protection 4. Respirators 5. Gloves 6. Masks	The following regulation addresses PPE: Occupational Safety and Health Administration (OSHA) 29 Code of Federal Regulations (CFR) Part 1910.132: Personal Protective Equipment.
EQUIPMENT PER TEAM TRANSPORTATION	1-Vehicle	Not Specified
PERSONNEL PER TEAM MANAGMENT AND OVERSIGHT	1 - National Incident Management System (NIMS) Type 1 Animal Emergency Response Team Leader	Not Specified
PERSONNEL PER TEAM MINIMUM	3	Not Specified
PERSONNEL PER TEAM SUPPORT	2 - Animal and Agriculture Damage Assessment Team Member	1. Animal and Agriculture Damage Assessment Team Member is not a NIMS typed position. 2. Team members are multidisciplinary and can include animal health professionals, agronomists, engineering specialists, logisticians, environmental experts, animal sheltering experts, communications specialists, and others.

FEMA

Resource Typing Definition for Situational Assessment
Animal Emergency Response

NOTES
Nationally typed resources represent the minimum criteria for the associated component and capability.

REFERENCES
1. FEMA, NIMS 509: Animal Emergency Response Team Leader
2. Occupational Safety and Health Administration (OSHA) 29 Code of Federal Regulations (CFR) Part 1910.132: Personal Protective Equipment, latest edition adopted

FEMA

Position Qualification for Public Health, Healthcare, and Emergency Medical Services
Animal Emergency Response

ANIMAL BEHAVIOR SPECIALIST

RESOURCE CATEGORY	Animal Emergency Response
RESOURCE KIND	Personnel
OVERALL FUNCTION	The Animal Behavior Specialist assesses behavioral status, identifies potentially dangerous behaviors, and makes behavioral recommendations for animals in one or more of the following competency areas during and after disasters: 1. Companion animals, including pets, service animals, and assistance animals 2. Livestock, including food or fiber animals and domesticated equine species 3. Wildlife animals, captive wildlife, and zoo animals 4. Laboratory animals
COMPOSITION AND ORDERING SPECIFICATIONS	1. This position can be ordered as a single resource. 2. Discuss logistics for deploying this position, such as security, lodging, transportation, and meals, prior to deployment 3. The position typically works 12 hours per shift, is self-sustainable for 72 hours, and is deployable for up to 14 days 4. Requestor specifies competency areas necessary based on the animal population the position will serve

Each type of resource builds on the qualifications of the type below it. For example, Type 1 qualifications include the qualifications in Type 2, plus an increase in capability. Type 1 is the highest qualification level.

COMPONENT	SINGLE TYPE	NOTES
DESCRIPTION	The Animal Behavior Specialist: 1. Triages animals to determine behavioral status 2. Classifies animals by behavioral status to determine the appropriate type of Animal Care and Handling Specialist or other staff to assign to handle them 3. Identifies potentially dangerous behaviors that could put the animal, the handler, or other people or animals at risk 4. Designs and coordinates animal enrichment and socialization programs 5. Develops and implements treatment plans to address animal behavioral issues for stray, abandoned, unattended, or relinquished animals 6. Consults on appropriate post-disaster animal disposition	Not Specified
EDUCATION	Not Specified	Not Specified
TRAINING	Completion of the following: 1. IS-100: Introduction to the Incident Command System, ICS-100 2. IS-200: Incident Command System for Single Resources and Initial Action Incidents 3. IS-700: National Incident Management System, An Introduction 4. IS-800: National Response Framework, An Introduction 5. Animal training, animal behavior training, or equivalent training	Not Specified

FEMA

Position Qualification for Public Health, Healthcare, and Emergency Medical Services
Animal Emergency Response

COMPONENT	SINGLE TYPE	NOTES
EXPERIENCE	Knowledge, Skills, and Abilities: 1. Extensive knowledge of species-typical and captive behavior of the animal populations represented 2. General understanding of animal welfare principles and processes 3. Extensive knowledge of animal training, motivation, and behavior modification principles 4. Practical experience working with the animal populations represented Experience: Three years of experience in applied animal behavior or animal training	Not Specified
PHYSICAL/MEDICAL FITNESS	1. Performs duties under moderate circumstances characterized by working consecutive 12-hour days under physical and emotional stress for sustained periods of time 2. Is able to work while wearing appropriate Personal Protective Equipment (PPE) 3. Keeps immunizations up to date and commensurate with mission	PPE is mission specific and may vary by work environment; it includes protective footwear, protective clothing for skin exposure, eye and ear protection, respirators, gloves, and masks.
CURRENCY	Routinely practices in a facility or field setting commensurate with the mission	Not Specified
PROFESSIONAL AND TECHNICAL LICENSES AND CERTIFICATIONS	Possesses one of the following certifications, or an equivalent: 1. Academy of Veterinary Behavior Technicians (AVBT) certification 2. American College of Veterinary Behaviorists (ACVB)-recognized certification 3. Animal Behavior Society: Associate Certified Applied Animal Behaviorist (ACAAB) 4. Animal Behavior Society: Certified Applied Animal Behaviorist (CAAB) 5. Certification Council for Professional Dog Trainers (CCPDT) certification 6. International Association of Animal Behavior Consultants (IAABC) certification	Not Specified

FEMA

Position Qualification for Public Health, Healthcare, and Emergency Medical Services
Animal Emergency Response

NOTES

Nationally typed resources represent the minimum criteria for the associated category.

REFERENCES

1. American College of Veterinary Behaviorists (ACVB), Board-Certified Veterinary Behaviorist
2. Animal Behavior Society, Certification Requirements and Application
3. Certification Council for Professional Dog Trainers (CCPDT), Dog Trainer Certification
4. CCPDT, Behavior Consultant Certification
5. FEMA, NIMS Guideline for the National Qualification System, November 2017
6. International Association of Animal Behavior Consultants (IAABC)

Position Qualification for Mass Care Services
Animal Emergency Response

ANIMAL CARE AND HANDLING SPECIALIST

RESOURCE CATEGORY	Animal Emergency Response
RESOURCE KIND	Personnel
OVERALL FUNCTION	The Animal Care and Handling Specialist provides proper care and handling of animals in one or more of the following competency areas: 1. Companion animals, including pets, service animals, and assistance animals 2. Livestock, including food or fiber animals and domesticated equine species 3. Wildlife animals, captive wildlife, and zoo animals 4. Laboratory animals
COMPOSITION AND ORDERING SPECIFICATIONS	1. This position can be ordered as a single resource or in conjunction with a NIMS typed team (such as an Animal Evacuation, Transport, and Re-Entry Team or an Animal Sheltering Team). 2. Discuss logistics for deploying this position, such as security, lodging, transportation, and meals, prior to deployment 3. This position typically works 12 hours per shift, is self-sustainable for 72 hours, and is deployable for up to 14 days 4. Requestor specifies competency areas necessary based on the animal population handled 5. For a Type 1 position, requestor specifies advanced specialty skill set(s) necessary

Each type of resource builds on the qualifications of the type below it. For example, Type 1 qualifications include the qualifications in Type 2, plus an increase in capability. Type 1 is the highest qualification level.

COMPONENT	TYPE 1	TYPE 2	NOTES
DESCRIPTION	Same as Type 2, PLUS: The Animal Care and Handling Specialist is capable of safe and humane handling of fractious, dangerous, or difficult-to-handle animals. These animals typically exhibit body language and behaviors consistent with fear, extreme submission, aggression, or attack directed at people or other animals. These animals may attempt to escape or resist capture or handling. In addition, uncastrated male livestock (including stallions, bulls, rams, and boars) may present a much-elevated level of handling hazard. Certain species of animals, such as non-human primates, big cats, and bears, are inherently dangerous and require specialized handling skills.	The Animal Care and Handling Specialist: 1. Assesses basic animal behavior 2. Provides daily animal care and containment 3. Ensures appropriate animal sanitation and biosecurity protocols are upheld 4. Contributes to animal census	Not Specified
EDUCATION	Not Specified	Not Specified	Not Specified

FEMA

Position Qualification for Mass Care Services
Animal Emergency Response

COMPONENT	TYPE 1	TYPE 2	NOTES
TRAINING	Same as Type 2	Completion of the following: 1. IS-100: Introduction to the Incident Command System, ICS-100 2. IS-200: Incident Command System for Single Resources and Initial Action Incidents 3. IS-700: National Incident Management System, An Introduction 4. IS-800: National Response Framework, An Introduction	Not Specified
EXPERIENCE	Same as Type 2, PLUS: Has a minimum of one year of regular, ongoing experience, with demonstrated ability to work with fractious, dangerous, or difficult-to-handle animals; required skills include: 1. Behavioral assessment 2. Capture and containment 3. Basic daily care 4. Restraint for procedures 5. Safe and humane handling	Routinely demonstrates the ability to work with non-fractious animals; required skills include: 1. Behavioral assessment 2. Capture and containment 3. Basic daily care 4. Restraint for procedures 5. Safe and humane handling	Specific relevant experience could include current and prior demonstrated competency as an Animal Control/Humane Officer, Veterinary Assistant, laboratory animal technician, zookeeper, or handler of dangerous animals at a livestock breeding facility.
PHYSICAL/MEDICAL FITNESS	Same as Type 2	1. Performs duties under arduous circumstances characterized by working consecutive 12-hour days under physical and emotional stress for sustained periods of time 2. Is able to work while wearing appropriate Personal Protective Equipment (PPE) 3. Keeps immunizations up to date and commensurate with mission	PPE is mission specific and may vary by working environment; it includes protective footwear, protective clothing for skin exposure, eye and ear protection, respirators, gloves, and masks
CURRENCY	Routinely performs animal care and handling duties with fractious, dangerous, or difficult-to-handle animals in a congregate animal housing facility related to the species listed; or service commensurate with the mission	Routinely performs animal care and handling duties in a congregate animal housing facility related to the species listed; or service commensurate with the mission	Not Specified
PROFESSIONAL AND TECHNICAL LICENSES AND CERTIFICATIONS	Not Specified	Not Specified	Not Specified

FEMA

Position Qualification for Mass Care Services
Animal Emergency Response

NOTES

Nationally typed resources represent the minimum criteria for the associated component and capability.

REFERENCES

1. FEMA, NIMS 508: Animal Transport Team – Companion Animal
2. FEMA, NIMS 508: Animal Transport Team – Livestock
3. FEMA, NIMS 508: Animal Sheltering Team – Cohabitated Shelter
4. FEMA, NIMS 508: Animal Sheltering Team – Collocated Shelter
5. FEMA, NIMS 508: Animal Sheltering Team – Animal-Only Shelter
6. FEMA, NIMS 508: Animal Evacuation and Re-entry Team
7. FEMA, NIMS 508: Animal Search and Rescue Team
8. FEMA, NIMS 509: Animal Control/Humane Officer
9. FEMA, NIMS 509: Veterinary Assistant
10. FEMA, NIMS Guideline for the National Qualification System, November 2017

FEMA

Position Qualification for On-scene Security, Protection and Law Enforcement
Animal Emergency Response

ANIMAL CONTROL/HUMANE OFFICER

RESOURCE CATEGORY	Animal Emergency Response
RESOURCE KIND	Personnel
OVERALL FUNCTION	The Animal Control/Humane Officer maintains public safety by enforcing animal-related laws and providing safe and humane capture and containment of animals in one or more of the following competency areas: 1. Companion animals, including pets, service animals, and assistance animals 2. Livestock, including food or fiber animals and domesticated equine species 3. Wildlife animals, captive wildlife, and zoo animals 4. Laboratory animals
COMPOSITION AND ORDERING SPECIFICATIONS	1. This position can be ordered as a single resource or in conjunction with a NIMS typed team (Animal Search and Rescue Team). 2. Discuss logistics for deploying this position, such as security, lodging, transportation, and meals, prior to deployment 3. This position typically works 12 hours per shift, is self-sustainable for 72 hours, and is deployable for up to 14 days 4. Requestor specifies competency areas necessary based on the animal population the position will serve 5. Requestor specifies specialty skills or certifications necessary, such as euthanasia or chemical capture 6. Requestor specifies whether Animal Control/Humane Officer should bring animal control vehicle and specialized equipment

Each type of resource builds on the qualifications of the type below it. For example, Type 1 qualifications include the qualifications in Type 2, plus an increase in capability. Type 1 is the highest qualification level.

COMPONENT	SINGLE TYPE	NOTES
DESCRIPTION	The Animal Control/Humane Officer: 1. Provides case management of animals involved in human bite cases 2. Provides safe and humane capture and containment of stray, abandoned, unattended, diseased, or injured animals 3. Assists with emergency euthanasia, as the Authority Having Jurisdiction (AHJ) authorizes 4. Ensures safety precautions for the public and animals when working in potentially dangerous situations with unfamiliar and unpredictable animals	Not Specified
EDUCATION	Not Specified	Not Specified
TRAINING	Completion of the following: 1. IS-100: Introduction to the Incident Command System 2. IS-200: Incident Command System for Single Resources and Initial Action Incidents 3. IS-700: National Incident Management System, An Introduction 4. IS-800: National Response Framework, An Introduction 5. National Animal Care & Control Association (NACA) certification training or equivalent formal training	Not Specified
EXPERIENCE	Two years of experience as an animal control officer, commensurate with the mission	Not Specified

FEMA

Position Qualification for On-scene Security, Protection and Law Enforcement
Animal Emergency Response

COMPONENT	SINGLE TYPE	NOTES
PHYSICAL/MEDICAL FITNESS	1. Performs duties under arduous circumstances characterized by working consecutive 12-hour days under physical and emotional stress for sustained periods of time 2. Is able to work while wearing appropriate Personal Protective Equipment (PPE) 3. Keeps immunizations up to date and commensurate with mission	PPE is mission specific and may vary by work environment; it includes protective footwear, protective clothing for skin exposure, eye and ear protection, respirators, gloves, and masks.
CURRENCY	Routinely provides animal control services commensurate with the mission	Not Specified
PROFESSIONAL AND TECHNICAL LICENSES AND CERTIFICATIONS	NACA certification or equivalent animal control certification	Not Specified

FEMA

Position Qualification for On-scene Security, Protection and Law Enforcement
Animal Emergency Response

NOTES
Nationally typed resources represent the minimum criteria for the associated component and capability.

REFERENCES
1. FEMA, NIMS 508: Animal Search and Rescue Team
2. FEMA, NIMS Guideline for the National Qualification System, November 2017
3. National Animal Care & Control Association (NACA) Guidelines

Position Qualification for Environmental Response/Health and Safety
Animal Emergency Response

ANIMAL DECONTAMINATION SPECIALIST

RESOURCE CATEGORY	Animal Emergency Response
RESOURCE KIND	Personnel
OVERALL FUNCTION	The Animal Decontamination Specialist provides a range of decontamination operations for animals in one or more of the following competency areas: 1. Companion animals, including pets, service animals, and assistance animals 2. Livestock, including food or fiber animals and domesticated equine species 3. Wildlife animals, captive wildlife, and zoo animals 4. Laboratory animals
COMPOSITION AND ORDERING SPECIFICATIONS	1. This position can be ordered as a single resource or in conjunction with a NIMS typed team (Companion Animal Decontamination Team). 2. Discuss logistics for deploying this position, such as security, lodging, transportation, and meals, prior to deployment 3. This position typically works 12 hours per shift, is self-sustainable for 72 hours, and is deployable for up to 14 days 4. Requestor can order this resource to augment the human decontamination lines of hazardous materials teams 5. Requestor specifies competency areas necessary based on the animal population the position will serve

Each type of resource builds on the qualifications of the type below it. For example, Type 1 qualifications include the qualifications in Type 2, plus an increase in capability. Type 1 is the highest qualification level.

COMPONENT	TYPE 1	TYPE 2	NOTES
DESCRIPTION	Same as Type 2, PLUS: 1. Handles and decontaminates animals for animal decontamination operations in nuclear or radiological incidents, including: 2. Removing radiological contaminants in a nuclear/radiological incident response	The Animal Decontamination Specialist: 1. Is a certified NIMS typed Animal Care and Handling Specialist, Animal Control/Humane Officer, Veterinarian, or Veterinary Assistant 2. Handles and decontaminates animals for animal decontamination operations other than nuclear or radiological; tasks include: a. Removing blood, urine, and feces from working dogs, working horses, or service animals b. Removing riot control agents, such as pepper spray, from working dogs, working horses, or service animals c. Removing contaminants associated with floodwater—such as debris, industrial chemicals, petroleum products, and biological pathogens—from animals after a disaster d. Cleaning oiled birds, mammals, and other wildlife e. Cleaning search-and-rescue cadaver dogs working in a contaminated environment f. Removing chemical contaminants in a chemical incident response g. Removing biotoxins in a biotoxin incident response	Not Specified
EDUCATION	Not Specified	Not Specified	Not Specified

FEMA

Position Qualification for Environmental Response/Health and Safety Animal Emergency Response

COMPONENT	TYPE 1	TYPE 2	NOTES
TRAINING	Same as Type 2, PLUS: 1. Training in accordance with the Occupational Safety and Health Administration (OSHA) 29 Code of Federal Regulations (CFR) Part 1910.1200: Hazard Communication Standard 2. Authority Having Jurisdiction (AHJ)-provided training in radiological responses and equipment use	Completion of the following: 1. Training for a NIMS typed Animal Care and Handling Specialist, Animal Control/Humane Officer, Veterinarian, or Veterinary Assistant 2. Training in accordance with the OSHA 29 CFR Part 1910.120: Hazardous Materials Awareness 3. AHJ-provided animal decontamination training	Not Specified
EXPERIENCE	Same as Type 2, PLUS: Participation in an animal radiological decontamination exercise	Participation in an animal decontamination exercise	Not Specified
PHYSICAL/MEDICAL FITNESS	Same as Type 2	1. Performs duties under arduous circumstances characterized by working consecutive 12-hour days under physical and emotional stress for sustained periods of time 2. Is able to work while wearing appropriate Personal Protective Equipment (PPE) 3. Keeps immunizations up to date and commensurate with mission	PPE is mission specific and may vary by working environment; it includes protective footwear, protective clothing for skin exposure, eye and ear protection, respirators, gloves, and masks.
CURRENCY	Same as Type 2	Participates in an animal decontamination exercise at least once every two years	Not Specified
PROFESSIONAL AND TECHNICAL LICENSES AND CERTIFICATIONS	Not Specified	Not Specified	Not Specified

FEMA

Position Qualification for Environmental Response/Health and Safety
Animal Emergency Response

NOTES

Nationally typed resources represent the minimum criteria for the associated component and capability.

REFERENCES

1. FEMA, NIMS 508: Companion Animal Decontamination Team
2. FEMA, NIMS 509: Animal Care and Handling Specialist
3. FEMA, NIMS 509: Animal Control/Humane Officer
4. FEMA, NIMS 509: Veterinarian
5. FEMA, NIMS 509: Veterinary Assistant
6. FEMA, NIMS Guideline for the National Qualification System, November 2017
7. Occupational Safety and Health Administration (OSHA) 29 Code of Federal Regulations (CFR) Part 1910.120: Hazardous Materials Awareness, latest edition adopted
8. OSHA 29 CFR Part 1910.1200: Hazard Communication Standard, latest edition adopted

Position Qualification for Environmental Response/Health and Safety
Animal Emergency Response

ANIMAL DEPOPULATION SPECIALIST

RESOURCE CATEGORY	Animal Emergency Response
RESOURCE KIND	Personnel
OVERALL FUNCTION	The Animal Depopulation Specialist depopulates animals when necessary because of public health and welfare concerns, disease exposure, threat, or infection. The Animal Depopulation Specialist has knowledge and expertise in one or more specialty areas of depopulation as outlined in the American Veterinary Medical Association (AVMA) Guidelines for the Euthanasia of Animals, and has knowledge, expertise, and experience in depopulating animals in one or more of the following competency areas: 1. Companion animals, including pets, service animals, and assistance animals 2. Livestock, including food or fiber animals and domesticated equine species 3. Wildlife animals, captive wildlife, and zoo animals 4. Laboratory animals
COMPOSITION AND ORDERING SPECIFICATIONS	1. This position can be ordered as a single resource or in conjunction with a NIMS typed team (Animal Depopulation Team). 2. Discuss logistics for deploying this position, such as security, lodging, transportation, and meals, prior to deployment 3. This position typically works 12 hours per shift, is self-sustainable for 72 hours, and is deployable for up to 14 days 4. Requestor discusses animal handling supplies and equipment necessary for specific species 5. Requestor specifies competency areas necessary based on the animal population the position will manage

Each type of resource builds on the qualifications of the type below it. For example, Type 1 qualifications include the qualifications in Type 2, plus an increase in capability. Type 1 is the highest qualification level.

COMPONENT	SINGLE TYPE	NOTES
DESCRIPTION	The Animal Depopulation Specialist: 1. Reviews the availability of animal handling facilities and evaluates their efficiency and safety (for animals and people) 2. Assesses the number, size, and weight of animals for depopulation 3. Assesses the staffing levels necessary for effective depopulation 4. Safely depopulates animals as trained, as expediently and humanely as possible 5. Ensures effective depopulation methodology for the species 6. Prepares animals for disposal	Not Specified
EDUCATION	Not Specified	Not Specified
TRAINING	Completion of the following: 1. IS-100: Introduction to the Incident Command System, ICS-100 2. IS-200: Incident Command System for Single Resources and Initial Action Incidents 3. IS-700: National Incident Management System, An Introduction 4. IS-800: National Response Framework, An Introduction 5. Authority Having Jurisdiction (AHJ)-approved training in the use of approved methods of euthanasia for different species	Not Specified
EXPERIENCE	Demonstrates to appropriate AHJ representative a skill set relevant to depopulation methodology, or has experience working on an Animal Depopulation Team	Not Specified

FEMA

Position Qualification for Environmental Response/Health and Safety
Animal Emergency Response

COMPONENT	SINGLE TYPE	NOTES
PHYSICAL/MEDICAL FITNESS	1. Performs duties under arduous circumstances characterized by working consecutive 12-hour days under physical and emotional stress for sustained periods of time 2. Is able to work while wearing appropriate Personal Protective Equipment (PPE) and commensurate with mission 3. Keeps immunizations up to date and commensurate with mission	PPE is mission specific and may vary by work environment; it includes protective footwear, protective clothing for skin exposure, eye and ear protection, respirators, gloves, and masks.
CURRENCY	Routinely trains with equipment and species commensurate with the mission	Not Specified
PROFESSIONAL AND TECHNICAL LICENSES AND CERTIFICATIONS	Not Specified	Not Specified

Position Qualification for Environmental Response/Health and Safety
Animal Emergency Response

NOTES

1. Nationally typed resources represent the minimum criteria for the associated component and capability.
2. Depopulation refers to the destruction of large numbers of animals in response to a public health or animal health emergency. Though animal welfare receives as much consideration as is practical, depopulation is sometimes necessary because of extenuating circumstances (Reference: APHIS 91-85-010, issued January 2016). Depopulation methods vary according to AHJ regulations. According to AVMA Guidelines for the Euthanasia of Animals, 2013, "Selection of the most appropriate method of euthanasia in any given situation depends on the species and number of animals involved, available means of animal restraint, skill of personnel, and other considerations."

REFERENCES

1. FEMA, NIMS 508: Animal Depopulation Team
2. FEMA, NIMS Guideline for the NQS, November 2017
3. American Veterinary Medical Association (AVMA) Guidelines for the Euthanasia of Animals, 2013
4. United States Department of Agriculture (USDA) Animal and Plant Health Inspection Service (APHIS), Highly Pathogenic Avian Influenza (HPAI), Depopulation and Disposal for Birds in Your HPAI-Infected Flock, APHIS 91-85-010, January 2016

Resource Typing Definition for Environmental Response/Health and Safety
Animal Emergency Response

ANIMAL DEPOPULATION TEAM

DESCRIPTION	The Animal Depopulation Team depopulates animals when necessary because of public health and welfare concerns, disease exposure, disease threat, or infection. This team may depopulate herds and flocks either in open areas or in confined spaces, using specialized equipment and protocols for the affected species.
RESOURCE CATEGORY	Animal Emergency Response
RESOURCE KIND	Team
OVERALL FUNCTION	This team: 1. Coordinates with Veterinary Medical Team, Companion Animal Decontamination Team, and disposal personnel as necessary 2. Has competency in one or more depopulation methods outlined in the American Veterinary Medical Association (AVMA) Guidelines for the Euthanasia of Animals 3. Works at existing facilities, such as poultry barns, sale barns, hospitals, and shelters, as well as in the field 4. Performs its duties for one or more of the following populations: a. Companion animals, including pets, service animals, and assistance animals b. Livestock, including food or fiber animals and domesticated equine species c. Wildlife, captive wildlife, or zoo animals d. Laboratory animals
COMPOSITION AND ORDERING SPECIFICATIONS	1. Discuss logistics for deploying this team, such as security, lodging, transportation, and meals, prior to deployment 2. This team typically works 12 hours per shift, is self-sustainable for 72 hours, and is deployable for up to 14 days 3. Requestor specifies type of species, number of existing animals, conditions under which team will operate, desired depopulation method, disease or welfare condition, complicating factors, and Personal Protective Equipment (PPE) requirements 4. Individuals responding with this team possess the knowledge and skills to perform depopulation activities humanely 5. Requestor specifies whether competency with particular animal populations is necessary 6. Discuss preparation supplies—such as water, foam, or carbon dioxide—as well as animal handling supplies and equipment necessary for species the team will manage 7. Requestor orders additional specialists separately, as necessary, such as Veterinarians, Veterinary Assistants, and Animal Care and Handling Specialists 8. Requestor orders separate resources to provide decontamination of personnel and equipment, if necessary 9. This team does not handle animal remains; requestor plans for disposal of animal remains separately in accordance with all applicable laws and regulations

Each type of resource builds on the qualifications of the type below it. For example, Type 1 qualifications include the qualifications in Type 2, plus an increase in capability. Type 1 is the highest qualification level.

COMPONENT	SINGLE TYPE	NOTES
EQUIPMENT PER TEAM GENERAL	1. Team transport vehicles (2) 2. Flashlights	Requestor orders additional trailer or vehicles, such as box trucks, if necessary, depending on species, conditions, presence or absence of disease, and handling requirements.
EQUIPMENT PER TEAM MEMBER COMMUNICATIONS	1. Two-way portable radio 2. Cell phone	Not Specified

FEMA

Resource Typing Definition for Environmental Response/Health and Safety
Animal Emergency Response

COMPONENT	SINGLE TYPE	NOTES
EQUIPMENT PER TEAM MEMBER PERSONAL PROTECTIVE EQUIPMENT	PPE is mission specific and may vary by work environment; it includes: 1. Protective footwear 2. Protective clothing for skin exposure 3. Eye and ear protection 4. Respirators 5. Gloves 6. Masks	The following regulation addresses PPE: Occupational Safety and Health Administration (OSHA) 29 Code of Federal Regulations (CFR) Part 1910.132: Personal Protective Equipment.
EQUIPMENT PER TEAM TECHNICAL	Equipment needs vary based on the conditions and species involved, and include: 1. Poultry foaming equipment 2. Poultry carbon dioxide equipment 3. Livestock captive bolt 4. Small animal injection materials	Not Specified
PERSONNEL PER TEAM MANAGEMENT AND OVERSIGHT	1 - National Incident Management System (NIMS) Type 1 Animal Emergency Response Team Leader	Not Specified
PERSONNEL PER TEAM MINIMUM	6	Not Specified
PERSONNEL PER TEAM SUPPORT	5 - NIMS Type 1 Animal Depopulation Specialist	Requestor orders additional Animal Care and Handling Specialists and Animal Control/Humane Officers as necessary.

FEMA

Resource Typing Definition for Environmental Response/Health and Safety
Animal Emergency Response

NOTES

1. Nationally typed resources represent the minimum criteria for the associated component and capability.
2. Depopulation refers to the destruction of large numbers of animals in response to a public health or animal health emergency. Though animal welfare receives as much consideration as is practical, depopulation is sometimes necessary because of extenuating circumstances (Reference: APHIS 91-85-010, issued January 2016). Depopulation methods vary according to Authority Having Jurisdiction (AHJ) regulations. According to AVMA Guidelines for the Euthanasia of Animals, 2013, "Selection of the most appropriate method of euthanasia in any given situation depends on the species and number of animals involved, available means of animal restraint, skill of personnel, and other considerations."

REFERENCES

1. FEMA, NIMS 508: Companion Animal Decontamination Team
2. FEMA, NIMS 508: Veterinary Team
3. FEMA, NIMS 509: Animal Care & Handling Specialist
4. FEMA, NIMS 509: Animal Control / Humane Officer
5. FEMA, NIMS 509: Animal Depopulation Specialist
6. FEMA, NIMS 509: Animal Emergency Response Team Leader
7. FEMA, NIMS 509: Veterinarian
8. FEMA, NIMS 509: Veterinary Assistant
9. American Veterinary Medical Association (AVMA) Guidelines for the Euthanasia of Animals, 2013
10. United States Department of Agriculture (USDA) Animal and Plant Health Inspection Service (APHIS), Highly Pathogenic Avian Influenza (HPAI), Depopulation and Disposal for Birds in Your HPAI-Infected Flock, APHIS 91-85-010, January 2016
11. Occupational Safety and Health Administration (OSHA) 29 Code of Federal Regulations (CFR) Part 1910.132: Personal Protective Equipment, latest edition adopted
12. National Wildfire Coordinating Group (NWCG), National Incident Management System Wildland Fire Qualification System Guide, PMS 310-1, Physical Fitness Levels, October 2016

Position Qualification for Mass Care Services
Animal Emergency Response

ANIMAL EMERGENCY RESPONSE SHELTER MANAGER

RESOURCE CATEGORY	Animal Emergency Response
RESOURCE KIND	Personnel
OVERALL FUNCTION	The Animal Emergency Response Shelter Manager provides leadership, supervision, and administrative support for the operation and demobilization of a temporary animal shelter in one or more of the following competency areas: 1. Companion animals, including pets, service animals, and assistance animals 2. Livestock, including food or fiber animals and domesticated equine species 3. Wildlife animals, captive wildlife, and zoo animals 4. Laboratory animals

COMPOSITION AND ORDERING SPECIFICATIONS	1. This position can be ordered as a single resource or in conjunction with a NIMS typed team (Animal Sheltering Team Animal Only Shelter, Animal Sheltering Team Cohabitated Shelter).
	2. Discuss logistics for deploying this position, such as security, lodging, transportation, and meals, prior to deployment
	3. This position typically works 12 hours per shift, is self-sustainable for 72 hours, and is deployable for up to 14 days
	4. Requestor specifies competency areas necessary based on the animal population the position will serve
	5. Requestor orders a Type 1 or Type 2 Animal Emergency Response Shelter Manager based on numerous factors, including:
	  a. Availability of established or identified buildings, utilities, ventilation, and supply lines
	  b. Availability of animal exercise areas
	  c. Environmental conditions
	  d. Feasibility of species colocation
	  e. Number of animals needing shelter

Each type of resource builds on the qualifications of the type below it. For example, Type 1 qualifications include the qualifications in Type 2, plus an increase in capability. Type 1 is the highest qualification level.

COMPONENT	TYPE 1	TYPE 2	NOTES
DESCRIPTION	Same as Type 2, PLUS: 1. Oversees setup, operations, and demobilization of a complex temporary animal shelter on a site without existing facilities 2. Determines site suitability for a temporary animal shelter, based on animal populations and species needing shelter 3. Designs a temporary animal shelter, including ventilation, electricity, water, exercise areas, and other facilities 4. Establishes a supply line for a temporary animal shelter 5. Equips and organizes a temporary animal shelter 6. Plans security for a temporary animal shelter 7. Plans for and sets up isolation, quarantine, and decontamination facilities, as necessary	1. Oversees setup, operations, and demobilization of a temporary animal shelter 2. Organizes and supervises all onsite staff and volunteers; delegate tasks, communicates instructions, enforces policies, and sets priorities 3. Manages information and facilitates communications to, from, and within the team 4. Ensures compliance with animal sheltering and animal welfare standards 5. Ensures facility and equipment maintenance, safety and biosecurity protocols, sanitation procedures, and general upkeep 6. Helps identify and allocate resources to support the shelter	Not Specified
EDUCATION	Not Specified	Not Specified	

Position Qualification for Mass Care Services
Animal Emergency Response

COMPONENT	TYPE 1	TYPE 2	NOTES
TRAINING	Same as Type 2	1. IS-100: Introduction to the Incident Command System, ICS-100 2. IS-200: Incident Command System for Single Resources and Initial Action Incidents 3. IS-300: Intermediate Incident Command System for Expanding Incidents 4. IS-700: National Incident Management System, An Introduction 5. IS-800: National Response Framework, An Introduction	Not Specified
EXPERIENCE	Same as Type 2, PLUS: Knowledge, Skills, and Abilities: 1. Knowledge of facility/site acquisition or identification 2. Knowledge of facility/site planning 3. Knowledge of facility renovation management Experience: Demonstrated experience in performing tasks required to set up a temporary animal shelter, including designing a shelter/facility layout and establishing organizational systems and workflow to ensure effective and efficient facility operations	Knowledge, Skills, and Abilities: Knowledge of facility management and maintenance Experience: Two years of experience managing an animal shelter or equivalent facility	Not Specified
PHYSICAL/MEDICAL FITNESS	Same as Type 2	1. Performs duties under arduous circumstances characterized by working consecutive 12-hour days under physical and emotional stress for sustained periods of time 2. Is able to work while wearing appropriate Personal Protective Equipment (PPE) 3. Keeps immunizations up to date and commensurate with mission	PPE is mission specific and may vary by work environment; it includes protective footwear, protective clothing for skin exposure, eye and ear protection, respirators, gloves, and masks.
CURRENCY	Functions in this position during an operational incident, exercise, drill, or simulation at least once every three years	Functions in this position during an operational incident, exercise, drill, or simulation at least once every five years	Not Specified
PROFESSIONAL AND TECHNICAL LICENSES AND CERTIFICATIONS	Not Specified	Not Specified	Not Specified

FEMA

Position Qualification for Mass Care Services
Animal Emergency Response

NOTES
Nationally typed resources represent the minimum criteria for the associated component and capability.

REFERENCES
1. FEMA, NIMS 508: Animal Sheltering Team – Animal-Only Shelter
2. FEMA, NIMS 508: Animal Sheltering Team – Cohabitated Shelter
3. FEMA, NIMS 508: Animal Sheltering Team – Collocated Shelter
4. FEMA, NIMS Guideline for the National Qualification System, November 2017
5. National Alliance of State Animal and Agricultural Emergency Programs (NASAAEP) Emergency Animal Sheltering Best Practices, September 2014
6. Association of Shelter Veterinarians (ASV) Guidelines for Standards of Care in Animal Shelters, 2010

Position Qualification for Operational Coordination
Animal Emergency Response

ANIMAL EMERGENCY RESPONSE TEAM LEADER

RESOURCE CATEGORY	Animal Emergency Response
RESOURCE KIND	Personnel
OVERALL FUNCTION	The Animal Emergency Response Team Leader leads and coordinates an Animal Emergency Response Team within the incident command structure, working in one or more of the following competency areas: 1. Companion animals, including pets, service animals, and assistance animals 2. Livestock, including food or fiber animals and domesticated equine species 3. Wildlife animals, captive wildlife, and zoo animals 4. Laboratory animals
COMPOSITION AND ORDERING SPECIFICATIONS	1. This position can be ordered as a single resource or in conjunction with a NIMS typed team (such as an Animal Evacuation, Transport, and Re-entry Team or an Animal Sheltering Team). 2. Discuss logistics for deploying this position, such as security, lodging, transportation, and meals, prior to deployment 3. This position typically works 12 hours per shift, is self-sustainable for 72 hours, and is deployable for up to 14 days 4. Requestor specifies competency areas necessary based on the animal population the position will serve

Each type of resource builds on the qualifications of the type below it. For example, Type 1 qualifications include the qualifications in Type 2, plus an increase in capability. Type 1 is the highest qualification level.

COMPONENT	SINGLE TYPE	NOTES
DESCRIPTION	The Animal Emergency Response Team Leader: 1. Serves as the team leader 2. Manages team members 3. Manages operational missions assigned to the team 4. Helps identify, procure, and allocate resources to support animal emergency response missions 5. Facilitates communications to, from, and within the team 6. Coordinates team activities within the incident command structure 7. Ensures team safety 8. Ensures compliance with animal welfare standards	Not Specified
EDUCATION	Not specified	Not Specified
TRAINING	1. IS-100: Introduction to the Incident Command System, ICS-100 2. IS-200: Incident Command System for Single Resources and Initial Action Incidents 3. ICS-300: Intermediate Incident Command System for Expanding Incidents 4. IS-700: National Incident Management System, An Introduction 5. IS-800: National Response Framework, An Introduction	
EXPERIENCE	1. Three years of experience as a supervisor or team leader in an animal care setting 2. Leadership experience in an animal emergency response incident or an operations-based exercise	Not Specified

FEMA

Position Qualification for Operational Coordination
Animal Emergency Response

COMPONENT	SINGLE TYPE	NOTES
PHYSICAL/MEDICAL FITNESS	1. Performs duties under moderate circumstances characterized by working consecutive 12-hour days under physical and emotional stress for sustained periods of time 2. Is able to work while wearing appropriate Personal Protective Equipment (PPE) 3. Keeps immunizations up to date and commensurate with mission	PPE is mission specific and may vary by work environment; it includes protective footwear, protective clothing for skin exposure, eye and ear protection, respirators, gloves, and masks.
CURRENCY	Functions in this position during an operational incident, exercise, drill, or simulation at least once every five years	Not Specified
PROFESSIONAL AND TECHNICAL LICENSES AND CERTIFICATIONS	Not Specified	Not Specified

FEMA

Position Qualification for Operational Coordination
Animal Emergency Response

NOTES

Nationally typed resources represent the minimum criteria for the associated component and capability.

REFERENCES

1. FEMA, NIMS 508: Animal Evacuation, Transport, and Re-entry Team
2. FEMA, NIMS 508: Animal Sheltering Team – Animal-Only Shelter
3. FEMA, NIMS 508: Animal Sheltering Team – Cohabitated Shelter
4. FEMA, NIMS 508: Animal Sheltering Team – Collocated Shelter
5. FEMA, NIMS Guideline for the National Qualification System, November 2017

FEMA

Resource Typing Definition for Critical Transportation
Animal Emergency Response

ANIMAL EVACUATION, TRANSPORT, AND RE-ENTRY TEAM

DESCRIPTION	The Animal Evacuation, Transport, and Re-entry Team evacuates animals from disaster areas to non-impacted areas by receiving animals from search-and-rescue teams, animal control personnel, shelters, or members of the community. This team coordinates the animals' return to the community, as appropriate. This team also transports animals from one location to another, as appropriate, to meet disaster response needs.
RESOURCE CATEGORY	Animal Emergency Response
RESOURCE KIND	Team
OVERALL FUNCTION	This team: 1. Coordinates with Animal Search and Rescue, Sheltering, and Veterinary Medical Teams, as necessary 2. Manages evacuation planning, activities, and transport 3. Identifies and documents animals 4. Initiates and continues animal tracking 5. Loads and unloads animals 6. Monitors animals throughout the evacuation, transport, and re-entry process 7. Transports or coordinates transport of evacuated animals to and from impacted areas 8. Reunifies or facilitates reunification of animals with their owners 9. Performs the above duties for one or more of the following populations: a. Companion animals, including pets, service animals, and assistance animals b. Livestock, including food or fiber animals and domesticated equine species c. Wildlife, captive wildlife, or zoo animals d. Laboratory animals
COMPOSITION AND ORDERING SPECIFICATIONS	1. Discuss logistics for deploying this team, such as security, lodging, transportation, and meals, prior to deployment 2. This team typically works 12 hours per shift, is self-sustainable for 72 hours, and is deployable for up to 14 days 3. Required personnel, supplies, and equipment depend on the operating environment (hot, warm, or cold), potential hazards, and type of mission (evacuation, re-entry, or both); requestor specifies additional training, certification, or experience required to operate in hazardous environments, if appropriate 4. Requestor specifies required competency with particular animal populations and tracking software/systems, as well as other evacuation, transport, and re-entry requirements 5. Requestor orders a Veterinarian or Veterinary Assistant to support team activities, if necessary 6. Discuss provision/acquisition of necessary animal identification and tracking supplies and equipment, such as GPS, microchips, universal microchip scanners, barcode systems and readers, ID supplies (tab band collars, livestock tags, or other ID suitable for species), intake forms, and request-for-rescue forms 7. Discuss provision of animal transport vehicles appropriate for the species transported, as well as the need for authorized drivers 8. Discuss the need for species-specific animal handling supplies and equipment 9. Requestor orders additional Animal Care and Handling Specialists or Animal Intake and Reunification Specialists as necessary, based on the incident

Each type of resource builds on the qualifications of the type below it. For example, Type 1 qualifications include the qualifications in Type 2, plus an increase in capability. Type 1 is the highest qualification level.

COMPONENT	TYPE 1	TYPE 2	TYPE 3	TYPE 4	NOTES
CAPACITY PER TEAM TRANSPORT	50 or more animals	31 to 49 animals	11 to 30 animals	1 to 10 animals	Capacity varies based on species transported.

FEMA

Resource Typing Definition for Critical Transportation
Animal Emergency Response

COMPONENT	TYPE 1	TYPE 2	TYPE 3	TYPE 4	NOTES
EQUIPMENT PER TEAM COMMUNICATIONS	Same as Type 2	Same as Type 3	Same as Type 4	1. Two-way portable radios as necessary for team communication 2. One cell phone per team member 3. One GPS per vehicle	Not Specified
EQUIPMENT PER TEAM MEMBER PERSONAL PROTECTIVE EQUIPMENT	Same as Type 2	Same as Type 3	Same as Type 4	PPE is mission specific and may vary by work environment; it includes: 1. Protective footwear 2. Protective clothing for skin exposure 3. Eye and ear protection 4. Respirators 5. Gloves 6. Masks	The following regulation addresses PPE: Occupational Safety and Health Administration (OSHA) 29 Code of Federal Regulations (CFR) Part 1910.132: Personal Protective Equipment.
EQUIPMENT PER TEAM TRANSPORT	Same as Type 2	Same as Type 3	Same as Type 4	Capture and containment equipment suitable for species transported	Not Specified
PERSONNEL PER TEAM MANAGEMENT AND OVERSIGHT	Same as Type 2	1 - National Incident Management System (NIMS) Type 1 Animal Emergency Response Team Leader	Not Specified	Not Specified	For Type 3 and Type 4 teams, a NIMS Type 1 Animal Emergency Response Team Leader may direct the team from a central dispatch location, or the existing chain of command may provide management and oversight, at the requestor's discretion.
PERSONNEL PER TEAM MINIMUM	12	9	4	2	Not Specified

FEMA

Resource Typing Definition for Critical Transportation
Animal Emergency Response

COMPONENT	TYPE 1	TYPE 2	TYPE 3	TYPE 4	NOTES
PERSONNEL PER TEAM SUPPORT	Same as Type 2, PLUS: 2 - NIMS Type 2 Animal Care and Handling Specialist 1 - NIMS Type 2 Animal Intake and Reunification Specialist	Same as Type 3, PLUS: 2 - NIMS Type 2 Animal Care and Handling Specialist 2 - NIMS Type 2 Animal Intake and Reunification Specialist	Same as Type 4, PLUS: 2 - NIMS Type 2 Animal Care and Handling Specialist	2 - NIMS Type 2 Animal Care and Handling Specialist	1. Add drivers based on the number of vehicles used for transport. Driver is not a NIMS typed position. 2. A driver may be one of the team members, depending on the vehicle and the team member's driving credentials. 3. Personnel listed represents the minimum number to accomplish the mission. Requestor will request additional personnel resources based on incident needs and the supported animal population.
TRANSPORTATION PER TEAM TRANSPORT	Same as Type 2	Same as Type 3	Same as Type 4	1. Number, type, and animal capacity of transport vehicles determined based on mission and species evacuated 2. Passenger vehicles, as appropriate for team mission	Not Specified

FEMA

Resource Typing Definition for Critical Transportation
Animal Emergency Response

NOTES

1. Nationally typed resources represent the minimum criteria for the associated component and capability.
2. Animal transport has inherent risks, particularly risks related to ventilation and temperature. Personnel must ensure safe and humane transport of animals, mitigating risks of animal injury or death in transit. Use special care in transporting very young, old, sick, or otherwise compromised animals. Regularly monitor the animals during transit.
3. See National Alliance of State Animal and Agricultural Emergency Programs (NASAAEP) Animal Evacuation and Transportation Best Practices for additional transportation guidelines.

REFERENCES

1. FEMA, NIMS 509: Animal Emergency Response Team Leader
2. FEMA, NIMS 509: Animal Care and Handling Specialist
3. FEMA, NIMS 509: Animal Intake and Reunification Specialist
4. FEMA, NIMS 509: Veterinarian
5. FEMA, NIMS 509: Veterinary Assistant
6. Occupational Safety and Health Administration (OSHA) 29 Code of Federal Regulations (CFR) Part 1910.132: Personal Protective Equipment, latest edition adopted
7. Animal Welfare Act and associated regulations, latest editions adopted
8. National Alliance of State Animal and Agricultural Emergency Programs (NASAAEP) Animal Evacuation and Transportation Best Practices, June 2012
9. NASAAEP Species Evacuation and Transport Guide, September 2014

Position Qualification for Mass Care Services
Animal Emergency Response

ANIMAL INTAKE AND REUNIFICATION SPECIALIST

RESOURCE CATEGORY	Animal Emergency Response
RESOURCE KIND	Personnel
OVERALL FUNCTION	The Animal Intake and Reunification Specialist provides animal intake, identification, tracking, and reunification (with owners or owner agents) during a disaster response. This is an administrative position that supports various animal emergency response teams in one or more of the following competency areas: 1. Companion animals, including pets, service animals, and assistance animals 2. Livestock, including food or fiber animals and domesticated equine species 3. Wildlife animals, captive wildlife, and zoo animals 4. Laboratory animals
COMPOSITION AND ORDERING SPECIFICATIONS	1. This position can be ordered as a single resource or in conjunction with a NIMS typed team (such as an Animal Sheltering Team or an Animal Evacuation, Transport, and Re-entry Team). 2. Discuss logistics for deploying this position, such as security, lodging, transportation, and meals, prior to deployment 3. This position typically works 12 hours per shift, is self-sustainable for 72 hours, and is deployable for up to 14 days 4. Requestor specifies competency areas necessary based on the animal population the position will serve

Each type of resource builds on the qualifications of the type below it. For example, Type 1 qualifications include the qualifications in Type 2, plus an increase in capability. Type 1 is the highest qualification level.

COMPONENT	TYPE 1	TYPE 2	NOTES
DESCRIPTION	Same as Type 2, PLUS: 1. Coordinates, oversees, and manages large and complex animal intake, identification, tracking, and reunification operations 2. Designs and implements intake, identification, tracking, and reunification protocols 3. Develops and maintains animal lost and found registry 4. Develops and implements just-in-time training on animal intake, identification, tracking, and reunification protocols	The Animal Intake and Reunification Specialist: 1. Identifies and documents animals 2. Tracks animal movements 3. Reunifies animals with owners or owner agents 4. Follows protocols for intake, identification, tracking, and reunification 5. Validates and reconciles animal inventory/census 6. Maintains animal lost and found registry 7. Reports validated animal counts and activities 8. Provides just-in-time training on animal intake, identification, tracking, and reunification protocols	Not Specified
EDUCATION	Not Specified	Not Specified	
TRAINING	Same as Type 2	Completion of the following: 1. IS-100: Introduction to the Incident Command System 2. IS-200: Incident Command System for Single Resources and Initial Action Incidents 3. IS-700: National Incident Management System, An Introduction 4. IS-800: National Response Framework, An Introduction	Not Specified

FEMA

Position Qualification for Mass Care Services
Animal Emergency Response

COMPONENT	TYPE 1	TYPE 2	NOTES
EXPERIENCE	Same as Type 2, PLUS: 1. Ability to design and implement animal tracking systems 2. Ability to design and implement animal intake, identification, and reunification protocols 3. Ability to develop just-in-time training	1. Working knowledge of animal species identification, animal breeds, and animal description (gender, color, size, and markings) 2. Ability to use animal management software 3. Ability to use social media to support animal lost and found registry	Not Specified
PHYSICAL/MEDICAL FITNESS	Same as Type 2	1. Performs duties under moderate circumstances characterized by working consecutive 12-hour days under physical and emotional stress for sustained periods of time 2. Is able to work while wearing appropriate Personal Protective Equipment (PPE) 3. Keeps immunizations up to date and commensurate with mission	PPE is mission specific and may vary by work environment; it includes protective footwear, protective clothing for skin exposure, eye and ear protection, respirators, gloves, and masks
CURRENCY	Routinely performs Type 1-level duties for the species listed, in conjunction with an animal control facility, animal shelter, veterinary hospital, livestock market, zoological institution, research facility, animal response organization, or related organization	Routinely performs Type 2-level duties for the species listed, in conjunction with an animal control facility, animal shelter, veterinary hospital, livestock market, zoological institution, research facility, animal response organization, or related organization	Not Specified
PROFESSIONAL AND TECHNICAL LICENSES AND CERTIFICATIONS	Not Specified	Not Specified	Not Specified

Position Qualification for Mass Care Services
Animal Emergency Response

NOTES

Nationally typed resources represent the minimum criteria for the associated component and capability.

REFERENCES

1. FEMA, NIMS 508: Animal Sheltering Team – Animal-Only Shelter
2. FEMA, NIMS 508: Animal Sheltering Team – Cohabitated Shelter
3. FEMA, NIMS 508: Animal Sheltering Team – Collocated Shelter
4. FEMA, NIMS 508: Animal Evacuation, Transport, and Re-entry Team
5. FEMA, NIMS Guideline for the National Qualification System, November 2017

FEMA

Resource Typing Definition for Mass Search and Rescue Operations
Animal Emergency Response

ANIMAL SEARCH AND RESCUE TEAM

DESCRIPTION	The Animal Search and Rescue Team locates, stabilizes, extricates, and evacuates animals in a disaster environment
RESOURCE CATEGORY	Animal Emergency Response
RESOURCE KIND	Team

OVERALL FUNCTION	This team:
	1. Coordinates and collaborates with other animal emergency response teams, as necessary
	2. Coordinates and plans animal search and rescue efforts
	3. Locates, captures, and contains displaced animals within the disaster zone
	4. Prepares animals for transport
	5. Identifies animals, documents rescue location, and records other relevant data to facilitate reunification with owners
	6. Coordinates live trapping and chemical capture as necessary
	7. Triages rescued animals for appropriate transport methods to prevent disease and mitigate medical and behavioral issues
	8. Performs the above duties for one or more of the following populations:
	  a. Companion animals, including pets, service animals, and assistance animals
	  b. Livestock, including food or fiber animals and domesticated equine species
	  c. Wildlife, captive wildlife, or zoo animals
	  d. Laboratory animals

COMPOSITION AND ORDERING SPECIFICATIONS	1. Discuss logistics for deploying this team, such as security, lodging, transportation, and meals, prior to deployment 2. This team typically works 12 hours per shift, is self-sustainable for 72 hours, and is deployable for up to 14 days 3. Discuss provision/acquisition of necessary supplies and equipment, such as restraint/control and capture equipment 4. Requestor specifies competency areas necessary based on animal population 5. Requestor orders additional Animal Search and Rescue Technician(s), Animal Control/Humane Officer(s), or Animal Care and Handling Specialist(s) to increase team capacity based on incident 6. Requestor makes provisions for veterinary care based on mission requirements

Each type of resource builds on the qualifications of the type below it. For example, Type 1 qualifications include the qualifications in Type 2, plus an increase in capability. Type 1 is the highest qualification level.

COMPONENT	TYPE 1	TYPE 2	TYPE 3	NOTES
CAPABILITY PER TEAM SEARCH AND RESCUE	Same as Type 2, PLUS: Can execute advanced rescue operations, such as high- and low-angle and swiftwater rescue	Same as Type 3, PLUS: Can perform water-based rescue and can use land- and water-based navigation technology to locate and geo-reference displaced animals	Land-based search and rescue only	Not Specified
CAPACITY PER TEAM SEARCH AND RESCUE	Responds to multiple jurisdictions, with multiple geographic areas, and manages approximately 50 companion animals or 25 livestock per 12-hour shift; may provide overall management and coordination remotely	Responds to a single jurisdiction with multiple geographic areas (or a single large, complex site) and manages approximately 24 animals per 12-hour shift	Responds to a single jurisdiction, manages a single area, and manages approximately 12 animals per 12-hour shift	Requestor may order additional single resources to augment team if necessary, based on number of animals.

FEMA

Resource Typing Definition for Mass Search and Rescue Operations
Animal Emergency Response

COMPONENT	TYPE 1	TYPE 2	TYPE 3	NOTES
EQUIPMENT PER TEAM ANIMAL HANDLING	Same as Type 2	Same as Type 3	Animal capture and handling equipment specific to the population served	Not Specified
EQUIPMENT PER TEAM ELECTRONIC	Same as Type 2, PLUS: Systems to input and transmit animal data to Animal Evacuation, Transport, and Re-Entry Team and Animal Sheltering Teams	Same as Type 3, PLUS: GPS	Digital cameras	Not Specified
EQUIPMENT PER TEAM MEMBER COMMUNICATIONS	Same as Type 2	Same as Type 3	1. Two-way portable radio 2. Cell phone	Not Specified
EQUIPMENT PER TEAM MEMBER PERSONAL PROTECTION EQUIPMENT	Same as Type 2	Same as Type 3	PPE is mission specific and may vary by work environment; it includes: 1. Protective footwear 2. Protective clothing for skin exposure 3. Eye and ear protection 4. Respirators 5. Gloves 6. Masks 7. Protective headgear	The following regulation addresses PPE: Occupational Safety and Health Administration (OSHA) 29 Code of Federal Regulations (CFR) Part 1910.132: Personal Protective Equipment.
PERSONNEL PER TEAM MANAGEMENT AND OVERSIGHT	Same as Type 2, PLUS: 4 - National Incident Management System (NIMS) Type 1 Animal Emergency Response Team Leader who is also a certified NIMS Type 3 Animal Search and Rescue Technician	1 - NIMS Type 1 Animal Emergency Response Team Leader who is also a certified NIMS Type 3 Animal Search and Rescue Technician	Not Specified	For Type 3 teams, any team member having completed ICS 300 may function as team leader, or the existing chain of command provides management and oversight at the requestor's discretion.
PERSONNEL PER TEAM MINIMUM	26	8	4	Not Specified
PERSONNEL PER TEAM SUPPORT	Same as Type 2, PLUS: 4 - NIMS Type 1 Animal Search and Rescue Technician 9 - NIMS Type 2 or Type 3 Animal Search and Rescue Technician 1 - NIMS Type 2 Veterinarian	7 - NIMS Type 2 Animal Search and Rescue Technician	1 - NIMS Type 3 Animal Search and Rescue Technician 3 - NIMS Type 2 Animal Care and Handling Specialist, NIMS Type 1 Animal Control/Humane Officer, or NIMS Type 3 Animal Search and Rescue Technician	1. A NIMS Type 2 Animal Care and Handling Specialist or NIMS Type 1 Animal Control/Humane Officer can replace a NIMS Type 3 Animal Search and Rescue Technician only on a Type 3 team. 2. Type 1 teams typically split into four subordinate teams.

FEMA

Resource Typing Definition for Mass Search and Rescue Operations
Animal Emergency Response

NOTES

Nationally typed resources represent the minimum criteria for the associated component and capability.

REFERENCES

1. FEMA, NIMS 508: Animal Evacuation, Transport, and Re-entry Team
2. FEMA, NIMS 508: Animal Sheltering Team – Animal-Only Shelter
3. FEMA, NIMS 508: Animal Sheltering Team – Cohabitated Shelter
4. FEMA, NIMS 508: Animal Sheltering Team – Collocated Shelter
5. FEMA, NIMS 508: Veterinary Medical Team
6. FEMA, NIMS 509: Animal Care and Handling Specialist
7. FEMA, NIMS 509: Animal Control/Humane Officer
8. FEMA, NIMS 509: Animal Emergency Response Team Leader
9. FEMA, NIMS 509: Animal Search and Rescue Technician
10. FEMA, NIMS 509: Veterinarian
11. Occupational Safety and Health Administration (OSHA) 29 Code of Federal Regulations (CFR) Part 1910.132: Personal Protective Equipment, latest edition adopted
12. National Alliance of State Animal and Agricultural Emergency Programs (NASAAEP) Best Practices for Animal Search and Rescue Training, latest published edition

Position Qualification for Mass Search and Rescue Operations
Animal Emergency Response

ANIMAL SEARCH AND RESCUE TECHNICIAN

RESOURCE CATEGORY	Animal Emergency Response
RESOURCE KIND	Personnel
OVERALL FUNCTION	The Animal Search and Rescue (ASAR) Technician locates, captures, contains, and evacuates animals in one or more of the following competency areas after a disaster: 1. Companion animals, including pets, service animals, and assistance animals 2. Livestock, including food or fiber animals and domesticated equine species 3. Wildlife animals, captive wildlife, and zoo animals 4. Laboratory animals
COMPOSITION AND ORDERING SPECIFICATIONS	1. This position can be ordered as a single resource or in conjunction with a NIMS typed team (Animal Search and Rescue Team). 2. Discuss logistics for deploying this position, such as security, lodging, transportation, and meals, prior to deployment 3. This position typically works 12 hours per shift, is self-sustainable for 72 hours, and is deployable for up to 14 days 4. Requestor specifies competency areas necessary based on the animal population the position will serve 5. Requestor specifies specialty areas (for example, fire, slack water, swiftwater) necessary based on incident requirements

Each type of resource builds on the qualifications of the type below it. For example, Type 1 qualifications include the qualifications in Type 2, plus an increase in capability. Type 1 is the highest qualification level.

COMPONENT	TYPE 1	TYPE 2	TYPE 3	NOTES
DESCRIPTION	Same as Type 2, PLUS: Performs safely in hazardous conditions involving animals exposed to a contaminated environment	Same as Type 3, PLUS: 1. Performs rescue in water-base situations 2. Performs technical rescue involving advanced animal rescue techniques and equipment, such as slack water and swiftwater rescue, high- and low-angle rescue, and use of slings and glides 3. Supports Urban Search and Rescue Team in compromised structure settings at an Operations Level	The Animal Search and Rescue Technician: 1. Performs rescue, rapid evacuation, and recovery in land-based situations 2. Performs rescue involving lifting and moving animals and equipment 3. Administers animal first aid 4. Ensures safety for self, team members, and operations by providing: a. Basic first aid for self and team members b. Simple decontamination of self and team members c. Ground vehicles for operations and support 5. Operates in environments with and without infrastructure, including areas affected by disasters and terrorism, and areas with compromised access to roadways, utilities, transportation, shelter, food, and water	Not Specified
EDUCATION	Not Specified	Not Specified	Not Specified	Not Specified

FEMA

Position Qualification for Mass Search and Rescue Operations
Animal Emergency Response

COMPONENT	TYPE 1	TYPE 2	TYPE 3	NOTES
TRAINING	AHJ ASAR Specialist Level training comparable to the National Alliance of State Animal and Agricultural Emergency Programs (NASAAEP) Best Practices for ASAR Training: Specialist Level	AHJ ASAR Technician Level training comparable to the NASAAEP Best Practices for ASAR Training: Technician Level	Authority Having Jurisdiction (AHJ) ASAR Operations Level training comparable to the NASAAEP Best Practices for ASAR Training: Operations Level	Not Specified
EXPERIENCE	Same as Type 2	Two years of experience in a disaster or emergency setting commensurate with the mission assignment	Not Specified	Not Specified
PHYSICAL/MEDICAL FITNESS	Same as Type 2	Same as Type 3	1. Performs duties under arduous circumstances characterized by working consecutive 12-hour days under physical and emotional stress for sustained periods of time 2. Is able to work while wearing appropriate Personal Protective Equipment (PPE) 3. Keeps immunizations up to date and commensurate with mission	PPE is mission specific and may vary by work environment; it includes protective footwear, protective clothing for skin exposure, eye and ear protection, respirators, gloves, and masks.
CURRENCY	Same as Type 2	Same as Type 3, PLUS: Routinely responds to ASAR requests at local, regional, and national levels	1.Responds to ASAR requests at local level 2. Participates in annual ASAR training exercise	Not Specified
PROFESSIONAL AND TECHNICAL LICENSES AND CERTIFICATIONS	AHJ-certified ASAR Specialist Level	AHJ-certified ASAR Technician Level	AHJ-certified ASAR Operations Level	Not Specified

FEMA

Position Qualification for Mass Search and Rescue Operations
Animal Emergency Response

NOTES
Nationally typed resources represent the minimum criteria for the associated component and capability.

REFERENCES
1. FEMA, NIMS 508: Animal Search and Rescue Team
2. FEMA, NIMS 508: Urban Search and Rescue Team
3. FEMA, NIMS Guideline for the NQS, November 2017
4. National Alliance of State Animal and Agricultural Emergency Programs (NASAAEP) Best Practices for Animal Search and Rescue Training, latest edition adopted
5. Occupational Safety and Health Administration (OSHA) 29 Code of Federal Regulations (CFR) Part 1910.132: Personal Protective Equipment, latest edition adopted

APPENDICES

211

Resource Typing Definition for Mass Care Services
Animal Emergency Response

ANIMAL SHELTERING TEAM - ANIMAL-ONLY SHELTER

DESCRIPTION	The Animal Sheltering Team - Animal-Only Shelter manages the oversight, setup, operations, communication, and demobilization of a temporary animal shelter. The team provides a safe and protected environment for displaced animal populations and meets their basic needs.
RESOURCE CATEGORY	Animal Emergency Response
RESOURCE KIND	Team

OVERALL FUNCTION	This team:
	1. Establishes and manages a temporary shelter for the safe and humane handling, care/husbandry, and housing of one of the following animal populations:
	  a. Companion animals, including pets, service animals, and assistance animals
	  b. Livestock, including food or fiber animals and domesticated equine species
	2. Meets animals' basic welfare needs
	3. Ensures proper animal identification, tracking, reunification, and reporting
	4. Coordinates with incident command; coordinates all facets of the animal response and intersecting components of the human response
	5. Maintains safety, biosecurity, and sanitation of the facility and equipment
	6. Provides appropriate security

COMPOSITION AND ORDERING SPECIFICATIONS	1. Discuss logistics for deploying this team, such as security, lodging, transportation, and meals, prior to deployment 2. This team works 12 hours per shift, is self-sustainable for 72 hours, and is deployable for up to 14 days 3. Requestor may order any combination of Veterinarian, Veterinary Assistant, and Veterinary Medical Team to support this team's activities, if necessary 4. Requestor may order a Shelter Facilities Support Team Leader to support shelter operations, if necessary 5. In an animal-only shelter, shelter staff provide full care of animals, which may be either stray (owner unknown) or temporarily away from their owners 6. Though owners may visit, they do not provide daily animal care 7. Average ordering ratio for Animal Care and Handling Specialists is one person to 15 animals; this ratio varies by species and shelter conditions 8. Requestor considers sheltering requirements for animals that are not medically or behaviorally suited for congregate sheltering 9. Requestor considers species-specific management needs

Each type of resource builds on the qualifications of the type below it. For example, Type 1 qualifications include the qualifications in Type 2, plus an increase in capability. Type 1 is the highest qualification level.

COMPONENT	TYPE 1	TYPE 2	TYPE 3	TYPE 4	NOTES
CAPACITY PER TEAM SHELTERING	Up to 500 animals	Up to 300 animals	Same as Type 4	Up to 100 animals	Not Specified
EQUIPMENT PER TEAM MEMBER COMMUNICATIONS	Same as Type 2	Same as Type 3	Same as Type 4	1.Two-way portable radio 2. Cell phone	Not Specified
EQUIPMENT PER TEAM MEMBER PERSONAL PROTECTIVE EQUIPMENT	Same as Type 2	Same as Type 3	Same as Type 4	PPE is mission specific and may vary be work environment; it includes: 1. Protective footwear 2. Protective clothing for skin exposure 3. Eye and ear protection 4. Gloves 5. Masks	The following regulation addresses PPE: Occupational Safety and Health Administration (OSHA) 29 Code of Federal Regulations (CFR) Part 1910.132: Personal Protective Equipment.

Resource Typing Definition for Mass Care Services Animal Emergency Response

COMPONENT	TYPE 1	TYPE 2	TYPE 3	TYPE 4	NOTES
EQUIPMENT PER TEAM SHELTERING	Same as Type 2	Same as Type 3	Requestor provides or obtains shelter kit appropriate for animal population served; kit may include: 1. Cages, crates, and other containment equipment 2. Collars, leashes, halters, lead ropes, lariat ropes 3. Animal ID supplies, such as collar tags 4. Muzzles 5. Food 6. Potable water 7. Bowls 8. Litter boxes 9. Litter 10. Cleaning and disinfecting supplies 11. Microchips/scanners 12. Vaccines 13. Animal intake forms	Not Specified	1. Team procures consumable animal supplies (food, litter, etc.) continually while shelter remains in operation. 2. Requestor specifies additional necessary equipment, including laptops, digital cameras, universal microchip scanners, intake forms, identification collars, or others.
LOCATION PER TEAM SHELTERING	Same as Type 2	Same as Type 3	A self-contained temporary shelter	Supplemental staff to support an established shelter	Not Specified
PERSONNEL PER TEAM MANAGEMENT AND OVERSIGHT	Same as Type 2, PLUS: 1 - National Incident Management System (NIMS) Type 1 Animal Emergency Response Team Leader 1 - Administrative Support	Same as Type 3, PLUS: 1 - NIMS Type 1 Animal Emergency Response Team Leader	Same as Type 4, PLUS: 1 - NIMS Type 2 Animal Emergency Response Shelter Manager	1 - NIMS Type 1 Animal Emergency Response Team Leader	1. Order one or more NIMS Type 1 Animal Emergency Response Shelter Manager if shelter planning, site selection, design, and setup are necessary to establish shelter. 2. Animal Emergency Response Team Leaders provide technical expertise for operations, planning, logistics, safety, finance, and administration of shelter operations. 3. Order additional Animal Emergency Response Team Leaders as appropriate based on scope and scale of shelter operations. 4. The Administrative Support position is not a NIMS typed position.
PERSONNEL PER TEAM MINIMUM	43	26	11	9	Not Specified

FEMA

Resource Typing Definition for Mass Care Services
Animal Emergency Response

COMPONENT	TYPE 1	TYPE 2	TYPE 3	TYPE 4	NOTES
PERSONNEL PER TEAM SUPPORT	Same as Type 2, PLUS: 13 - NIMS Type 2 Animal Care and Handling Specialist 2 - Animal Intake and Reunification Specialist	Same as Type 3, PLUS: 13 - NIMS Type 2 Animal Care and Handling Specialist 1 - NIMS Type 2 Animal Intake and Reunification Specialist	Same as Type 4, PLUS: 1 - NIMS Type 2 Animal Intake and Reunification Specialist	2 - NIMS Type 1 Animal Care and Handling Specialist 5 - NIMS Type 2 Animal Care and Handling Specialist 1 - NIMS Type 2 Animal Intake and Reunification Specialist	Order additional Animal Care and Handling Specialists and Animal Intake and Reunification Specialists as appropriate based on scope and scale of shelter operations; for example: 1. To help with animal intake as shelter opens 2. If operating hours exceed 12 hours per day 3. If number of animals in shelter exceeds 500

FEMA

Resource Typing Definition for Mass Care Services
Animal Emergency Response

NOTES

Nationally typed resources represent the minimum criteria for the associated component and capability.

REFERENCES

1. FEMA, NIMS 509: Animal Emergency Response Team Leader
2. FEMA, NIMS 509: Animal Emergency Response Shelter Manager
3. FEMA, NIMS 509: Animal Care and Handling Specialist
4. FEMA, NIMS 509: Animal Intake and Reunification Specialist
5. FEMA, NIMS 509: Veterinarian
6. FEMA, NIMS 509: Veterinary Assistant
7. FEMA, NIMS 508: Veterinary Medical Team
8. FEMA, NIMS 509: Shelter Facilities Support Team Leader
9. Occupational Safety and Health Administration (OSHA) 29 Code of Federal Regulations (CFR) Part 1910.132: Personal Protective Equipment, latest edition adopted
10. National Alliance of State Animal and Agricultural Emergency Programs (NASAAEP) Emergency Animal Sheltering Best Practices, September 2014

FEMA

Resource Typing Definition for Mass Care Services
Animal Emergency Response

ANIMAL SHELTERING TEAM - COHABITATED SHELTER

DESCRIPTION	The Animal Sheltering Team - Cohabitated Shelter manages the oversight, setup, operations, communication, and demobilization of a temporary animal shelter. The team provides a safe and protected environment for displaced animal populations and meets their basic needs.
RESOURCE CATEGORY	Animal Emergency Response
RESOURCE KIND	Team

OVERALL FUNCTION	This team:
	1. Establishes and manages a temporary shelter for the safe and humane handling, care/husbandry, and housing of one of the following animal populations:
	  a. Companion animals, including pets, service animals, and assistance animals
	  b. Livestock, including food or fiber animals and domesticated equine species
	2. Meets animals' basic welfare needs
	3. Ensures proper animal identification, tracking, reunification, and reporting
	4. Coordinates with incident command; coordinates all facets of the animal response and intersecting components of the human response
	5. Maintains safety, biosecurity, and sanitation of the facility and equipment
	6. Provides appropriate security

COMPOSITION AND ORDERING SPECIFICATIONS	1. Discuss logistics for deploying this team, such as security, lodging, transportation, and meals, prior to deployment 2. This team works 12 hours per shift, is self-sustainable for 72 hours, and is deployable for up to 14 days 3. Requestor may order any combination of Veterinarian, Veterinary Assistant, and Veterinary Medical Team to support this team's activities, if necessary 4. Requestor may order a Shelter Facilities Support Team Leader to support shelter operations, if necessary 5. Cohabitated shelters place animals and owners in the same space 6. Animal owners provide primary care for their own animals with their own supplies, which the sheltering entity may supplement if necessary and if available 7. Requestor considers sheltering requirements for animals that are not medically or behaviorally suited for congregate sheltering 8. Requestor considers species-specific management needs

Each type of resource builds on the qualifications of the type below it. For example, Type 1 qualifications include the qualifications in Type 2, plus an increase in capability. Type 1 is the highest qualification level.

COMPONENT	TYPE 1	TYPE 2	TYPE 3	TYPE 4	NOTES
CAPACITY PER TEAM SHELTERING	Up to 500 animals	Up to 300 animals	Same as Type 4	Up to 100 animals	Not Specified
EQUIPMENT PER TEAM COMMUNICATIONS	Same as Type 2	Same as Type 3	Same as Type 4	1. Two-way portable radio 2. Cell phone	Not Specified
EQUIPMENT PER TEAM MEMBER PERSONAL PROTECTIVE EQUIPMENT	Same as Type 2	Same as Type 3	Same as Type 4	PPE is mission specific and may vary by work environment; it includes: 1. Protective footwear 2. Protective clothing for skin exposure 3. Eye and ear protection 4. Gloves 5. Mask	The following regulation addresses PPE: Occupational Safety and Health Administration (OSHA) 29 Code of Federal Regulations (CFR) Part 1910.132: Personal Protective Equipment.

Resource Typing Definition for Mass Care Services
Animal Emergency Response

COMPONENT	TYPE 1	TYPE 2	TYPE 3	TYPE 4	NOTES
EQUIPMENT PER TEAM SHELTERING	Same as Type 2	Same as Type 3	Requestor provides or obtains shelter kit appropriate for animal population served; kit may include: 1. Cages, crates, and other containment equipment 2. Collars, leashes, halters, lead ropes, lariat ropes 3. Animal ID supplies, such as collar tags 4. Muzzles 5. Food 6. Potable water 7. Bowls 8. Litter boxes 9. Litter 10. Cleaning and disinfecting supplies 11. Microchips/scanners 12. Vaccines 13. Animal intake forms	Not Specified	1. Team procures consumable animal supplies (food, litter, and so on) continually while shelter remains in operation. 2. Requestor specifies additional necessary equipment, including laptop computers, digital cameras, universal microchip scanners, intake forms, or identification collars.
LOCATION PER TEAM SHELTERING	Same as Type 2	Same as Type 3	A self-contained temporary shelter	Supplemental staff to support an established shelter	Not Specified
PERSONNEL PER TEAM MANAGEMENT AND OVERSIGHT	Same as Type 2, PLUS: 1 - National Incident Management System (NIMS) Type 1 Animal Emergency Response Team Leader 1 - Administrative Support	Same as Type 3, PLUS: 1 - NIMS Type 1 Animal Emergency Response Team Leader	Same as Type 4, PLUS: 1 - NIMS Type 2 Animal Emergency Response Shelter Manager	1 - NIMS Type 1 Animal Emergency Response Team Leader	1. Order one or more NIMS Type 1 Animal Emergency Response Shelter Manager if shelter planning, site selection, design, and setup are necessary to establish shelter. 2. Animal Emergency Response Team Leaders provide technical expertise for operations, planning, logistics, safety, finance, and administration of shelter operations. 3. Order additional Animal Emergency Response Team Leaders as appropriate based on scope and scale of shelter operations. 4. The Administrative Support position is not a NIMS typed position.
PERSONNEL PER TEAM MINIMUM	10	7	5	4	Not Specified

FEMA

Resource Typing Definition for Mass Care Services
Animal Emergency Response

COMPONENT	TYPE 1	TYPE 2	TYPE 3	TYPE 4	NOTES
SUPPORT PER TEAM SHELTERING	Sheltering Same as Type 2, PLUS: 1 - NIMS Type 2 Animal Intake and Reunification Specialist	Same as Type 3, PLUS: 1 - NIMS Type 2 Animal Intake and Reunification Specialist	Same as Type 4	2 - NIMS Type 1 Animal Care and Handling Specialist 1 - NIMS Type 2 Animal Intake and Reunification Specialist	Order additional Animal Care and Handling Specialists and Animal Intake and Reunification Specialists as appropriate based on scope and scale of shelter operations; for example: 1. To help with animal intake as shelter opens 2. If operating hours exceed 12 hours per day 3. If number of animals in shelter exceeds 500 4. If shelter includes both animals that owners care for and animals that are unattended

FEMA

Resource Typing Definition for Mass Care Services
Animal Emergency Response

NOTES
Nationally typed resources represent the minimum criteria for the associated component and capability.

REFERENCES

1. FEMA, NIMS 508: Veterinary Medical Team
2. FEMA, NIMS 509: Animal Emergency Response Shelter Manager
3. FEMA, NIMS 509: Animal Emergency Response Team Leader
4. FEMA, NIMS 509: Animal Care and Handling Specialist
5. FEMA, NIMS 509: Animal Intake and Reunification Specialist
6. FEMA, NIMS 509: Shelter Facilities Support Team Leader
7. FEMA, NIMS 509: Veterinarian
8. FEMA, NIMS 509: Veterinary Assistant
9. Occupational Safety and Health Administration (OSHA) 29 Code of Federal Regulations (CFR) Part 1910.132: Personal Protective Equipment, latest edition adopted
10. National Alliance of State Animal and Agricultural Emergency Programs (NASAAEP) Emergency Animal Sheltering Best Practices, September 2014

Resource Typing Definition for Mass Care Services
Animal Emergency Response

ANIMAL SHELTERING TEAM - COLLOCATED SHELTER

DESCRIPTION	The Animal Sheltering Team - Collocated Shelter manages the oversight, setup, operations, communication, and demobilization of a temporary animal shelter. The team provides a safe and protected environment for displaced animal populations and meets their basic needs.
RESOURCE CATEGORY	Animal Emergency Response
RESOURCE KIND	Team

OVERALL FUNCTION	This team:
	1. Establishes and manages a temporary shelter for the safe and humane handling, care/husbandry, and housing of one of the following animal populations:
	  a. Companion animals, including pets, service animals, and assistance animals
	  b. Livestock, including food or fiber animals and domesticated equine species
	2. Meets animals' basic welfare needs
	3. Ensures proper animal identification, tracking, reunification, and reporting
	4. Coordinates with incident command; coordinates all facets of the animal response and intersecting components of the human response
	5. Maintains safety, biosecurity, and sanitation of the facility and equipment
	6. Provides appropriate security

COMPOSITION AND ORDERING SPECIFICATIONS	1. Discuss logistics for deploying this team, such as security, lodging, transportation, and meals, prior to deployment 2. This team works 12 hours per shift, is self-sustainable for 72 hours, is deployable for up to 14 days 3. Requestor may order any combination of Veterinarian, Veterinary Assistant, and Veterinary Medical Team to support this team's activities, if necessary 4. Requestor may order a Shelter Facilities Support Team Leader to support shelter operations, if necessary 5. Collocated shelters place animals within local commuting distance of their owners 6. Animal owners provide basic daily care for their animals in the collocated shelter 7. Requestor considers sheltering requirements for animals that are not medically or behaviorally suited for congregate sheltering 8. Requestor considers species-specific management needs

Each type of resource builds on the qualifications of the type below it. For example, Type 1 qualifications include the qualifications in Type 2, plus an increase in capability. Type 1 is the highest qualification level.

COMPONENT	TYPE 1	TYPE 2	TYPE 3	TYPE 4	NOTES
CAPACITY PER TEAM SHELTERING	Up to 500 animals	Up to 300 animals	Same as Type 4	Up to 100 animals	Not Specified
EQUIPMENT PER TEAM MEMBER COMMUNICATIONS	Same as Type 2	Same as Type 3	Same as Type 4	1. Two-way portable radio 2. Cell phone	Not Specified
EQUIPMENT PER TEAM MEMBER PERSONAL PROTECTIVE EQUIPMENT	Same as Type 2	Same as Type 3	Same as Type 4	PPE is mission specific and may vary by work environment; it includes: 1. Protective footwear 2. Protective clothing for skin exposure 3. Eye and ear protection 4. Gloves 5. Masks	The following regulation addresses PPE: Occupational Safety and Health Administration (OSHA) 29 Code of Federal Regulations (CFR) Part 1910.132: Personal Protective Equipment.

Resource Typing Definition for Mass Care Services
Animal Emergency Response

COMPONENT	TYPE 1	TYPE 2	TYPE 3	TYPE 4	NOTES
EQUIPMENT PER TEAM SHELTERING	Same as Type 2	Same as Type 3	Requestor provides or obtains shelter kit appropriate for animal population served; kit may include: 1. Cages, crates, and other containment equipment 2. Collars, leashes, halters, lead ropes, lariat ropes 3. Animal ID supplies, such as collar tags 4. Muzzles 5. Food 6. Potable water 7. Bowls 8. Litter boxes 9. Litter 10. Cleaning and disinfecting supplies 11. Microchips/scanners 12. Vaccines 13. Animal intake forms	Not Specified	1. Team procures consumable animal supplies (food, litter, and so on) continually while shelter remains in operation. 2. Requestor specifies additional necessary equipment, including laptop computers, digital cameras, universal microchip scanners, intake forms, or identification collars.
LOCATION PER TEAM SHELTERING	Same as Type 2	Same as Type 3	A self-contained temporary shelter	Supplemental staff to support an established shelter	Not Specified
PERSONNEL PER TEAM MANAGEMENT AND OVERSIGHT	Same as Type 2, PLUS: 1 - National Incident Management System (NIMS) Type 1 Animal Emergency Response Team Leader 1 - Administrative Support	Same as Type 3, PLUS: 1 - NIMS Type 1 Animal Emergency Response Team Leader	Same as Type 4, PLUS: 1 - NIMS Type 2 Animal Emergency Response Shelter Manager	1 - NIMS Type 1 Animal Emergency Response Team Leader	1. Order one or more NIMS Type 1 Animal Emergency Response Shelter Manager if shelter planning, site selection, design, and setup are necessary to establish shelter. 2. Animal Emergency Response Team Leaders provide technical expertise for operations, planning, logistics, safety, finance, and administration of shelter operations. 3. Order additional Animal Emergency Response Team Leaders as appropriate based on scope and scale of shelter operations. 4. The Administrative Support position is not a NIMS typed position.
PERSONNEL PER TEAM MINIMUM	15	11	8	7	Not Specified

FEMA

Resource Typing Definition for Mass Care Services
Animal Emergency Response

COMPONENT	TYPE 1	TYPE 2	TYPE 3	TYPE 4	NOTES
PERSONNEL PER TEAM SUPPORT	Same as Type 2, PLUS: 1 - NIMS Type 2 Animal Care and Handling Specialist 1 - Animal Intake and Reunification Specialist	Same as Type 3, PLUS: 1 - NIMS Type 2 Animal Care and Handling Specialist 1 - Animal Intake and Reunification Specialist	Same as Type 4	2 - NIMS Type 1 Animal Care and Handling Specialist 3 - NIMS Type 2 Animal Care and Handling Specialist 1 - Animal Intake and Reunification Specialist	Order additional Animal Care and Handling Specialists and Animal Intake and Reunification Specialists as appropriate based on scope and scale of shelter operations; for example: 1. To help with animal intake as shelter opens 2. If operating hours exceed 12 hours per day 3. If number of animals in shelter exceeds 500 4. If shelter includes both animals that owners care for and animals that are unattended

FEMA

Resource Typing Definition for Mass Care Services
Animal Emergency Response

NOTES
Nationally typed resources represent the minimum criteria for the associated component and capability.

REFERENCES

1. FEMA, NIMS 508: Veterinary Medical Team
2. FEMA, NIMS 509: Animal Emergency Response Shelter Manager
3. FEMA, NIMS 509: Animal Emergency Response Team Leader
4. FEMA, NIMS 509: Animal Care and Handling Specialist
5. FEMA, NIMS 509: Animal Intake and Reunification Specialist
6. FEMA, NIMS 509: Shelter Facilities Support Team Leader
7. FEMA, NIMS 509: Veterinarian
8. FEMA, NIMS 509: Veterinary Assistant
9. Occupational Safety and Health Administration (OSHA) 29 Code of Federal Regulations (CFR) Part 1910.132: Personal Protective Equipment, latest edition adopted
10. National Alliance of State Animal and Agricultural Emergency Programs (NASAAEP) Emergency Animal Sheltering Best Practices, September 2014

Resource Typing Definition for Environmental Response/Health and Safety
Animal Emergency Response

COMPANION ANIMAL DECONTAMINATION TEAM

DESCRIPTION	A Companion Animal Decontamination Team provides companion animal intake, radiological assessment (if applicable), decontamination, and re-monitoring (if applicable)—followed by release to owners if animals are adequately clean or transfer to a longer-term holding facility if animals are persistently contaminated. For purposes of this document only, companion animals include pets, assistance animals, and service animals.
RESOURCE CATEGORY	Animal Emergency Response
RESOURCE KIND	Team
OVERALL FUNCTION	This team manages the decontamination of companion animals after incidents involving hazardous materials, including debris, floodwaters, and radiological contamination. Specifically, this team: 1. Sets up all equipment at a designated "warm zone" site 2. Accepts animals from their owners or caretakers for rapid triage (behavioral and health), identification, and initial monitoring; Note: In some radiological incidents, this team may ask owners to participate actively in decontaminating their animals, per incident policies; Owner participation is less likely in non-radiological incidents 3. Transfers severely injured or ill animals to veterinary medical personnel for stabilization prior to decontamination, if resources permit, according to incident policies 4. Decontaminates animals using techniques appropriate for contaminant, species, breed, environmental conditions, and available resources 5. Monitors and re-monitors animals after decontamination in radiological incidents 6. Releases adequately decontaminated animals to their owners, or transfers animals to emergency animal shelters if owners are not available to claim animals 7. Performs repeat decontamination on persistently contaminated animals, or transfers such animals to a designated holding area 8. Disposes of wastewater and other waste in accordance with incident policies
COMPOSITION AND ORDERING SPECIFICATIONS	1. Discuss logistics for deploying this team, such as security, lodging, transportation, and meals, prior to deployment 2. This team typically works 12 hours per shift under ideal environmental conditions; challenging environments (heat, cold, and precipitation) could reduce the team's ability to safely work continuously for a full shift 3. This team is self-sustainable for 24 hours and is deployable for up to 14 days 4. Requestor specifies desired capabilities for a single line of animal decontamination operations for a specific period 5. This team may require supporting utilities, including water, power, and sewer/wastewater disposal. For large-scale incidents, this team may require additional supplies, including Personal Protective Equipment (PPE), disposable leashes, pet collars, surfactant, drying towels, and other supplies. Requestor should discuss expected numbers of animals for decontamination and team supply needs before placing order 6. Particularly in radiological incidents, requestor should discuss whether this team will establish a second or even third decontamination line in which owners can bathe and dry their own animals, with monitoring, coaching, and supervision. Requestor may need to order additional single resources to support this option. Consider local jurisdictional policies and liability in decisions about owners decontaminating their own animals 7. Discuss whether the incident requires 24-hour operations and whether very large numbers of animals will need monitoring or decontamination, and order teams accordingly 8. Discuss any additional just-in-time training necessary to comply with incident policy

Each type of resource builds on the qualifications of the type below it. For example, Type 1 qualifications include the qualifications in Type 2, plus an increase in capability. Type 1 is the highest qualification level.

FEMA

Resource Typing Definition for Environmental Response/Health and Safety
Animal Emergency Response

COMPONENT	TYPE 1	TYPE 2	TYPE 3	TYPE 4	NOTES
CAPACITY PER TEAM MINIMUM	Radiological decontamination: 1. Team can sustain one line of animals moving from station to station, with different personnel at each station 2. Minimum capacity of an estimated 10-12 animals per hour per line, with periodic rotation of some personnel 3. Capacity depends on environmental temperature and conditions 4. Due to PPE, ergonomic, and environmental challenges, team should include three rotating shifts of personnel to sustain operations continuously for 12 hours 5. Team may be able to sustain a second or even a third line, including line(s) for owners who volunteer to decontaminate their own animals, depending on jurisdictional and incident policies 6. Total minimum capacity estimated at 144 animals per shift, per line	Radiological decontamination: 1. Minimum capacity of an estimated 5 animals per hour with periodic rotation of some personnel 2. Capacity depends on environmental temperatures and conditions 3. Team may be able to sustain a second line for owners who volunteer to decontaminate their own animals, depending on jurisdictional and incident policies 4. Total minimum capacity of up to 50 animals per shift, not including pets decontaminated by owners	Floodwater or debris decontamination: 1. Single line of decontamination with a capacity of approximately 4-6 animals per hour or 50 animals per day	Not Specified	For radiological incidents, not every animal will need decontamination. Cited numbers emphasize animals that need decontamination and not those that the team simply returns to their owners.

FEMA

Resource Typing Definition for Environmental Response/Health and Safety
Animal Emergency Response

COMPONENT	TYPE 1	TYPE 2	TYPE 3	TYPE 4	NOTES
EQUIPMENT PER TEAM DECONTAMINATION	Same as Type 2, PLUS: 1. One additional set of decontamination pools and tables 2. Two additional radiation monitoring devices (dosimeters)	1. 500-gallon water tank 2. Water heater capable of supporting two lines of decontamination 3. Water hoses and sprayers 4. Shallow decontamination pools (2) 5. Elevated animal handling tables for use inside pools (2) 6. Drain hoses or sump pumps 7. Tents (3-4) for intake area, animal holding area, and responder recovery area 8. Powered air purifying respirators or filter respirators 9. Personal dosimeter for each team member 10. Handheld radiation monitors (minimum 2) 11. Power cords and surge protectors 12. Veterinary diagnostic equipment (stethoscope, light, thermometers, etc.) 13. Animal cages for dogs and cats 14. Generator and fuel capable of supporting operations for a minimum of 24 hours	1. 250-gallon water tank 2. Water heater capable of supporting a single line of decontamination 3. Water hoses and sprayers 4. Shallow decontamination pool 5. Elevated animal handling table for use inside pool 6. Drain hoses or sump pumps 7. Tents or canopies (2-3) for intake area, animal holding area, and responder recovery area 8. Animal cages for dogs and cats 9. Generator and fuel capable of supporting operations for a minimum of 24 hours	Not Specified	1. Requestor and provider discuss equipment needs and what supplies the requestor will provide. 2. If location has running water or requestor can provide water, requestor may waive water requirements
EQUIPMENT PER TEAM MEMBER COMMUNICATION	Same as Type 2	Same as Type 3	Cell phone	Not Specified	Not Specified
EQUIPMENT PER TEAM MEMBER PERSONAL PROTECTIVE EQUIPMENT	Same as Type 2	Same as Type 3, PLUS: 1. Individual dosimeters for every team member 2. N-100 or better respirator	PPE is mission specific and may vary by work environment; it includes: 1. Protective footwear 2. Protective clothing for skin exposure 3. Eye and ear protection 4. Respirators 5. Gloves 6. Masks	Not Specified	The following regulation addresses PPE: Occupational Safety and Health Administration (OSHA) 29 Code of Federal Regulations (CFR) Part 1910.132: Personal Protective Equipment.

FEMA

Resource Typing Definition for Environmental Response/Health and Safety
Animal Emergency Response

COMPONENT	TYPE 1	TYPE 2	TYPE 3	TYPE 4	NOTES
PERSONNEL PER TEAM MANAGEMENT AND OVERSIGHT	Same as Type 2, PLUS: 1 - National Incident Management System (NIMS) Type 1 Animal Emergency Response Team Leader (deputy)	Same as Type 3	1 - NIMS Type 1 Animal Emergency Response Team Leader	Not Specified	Not Specified
PERSONNEL PER TEAM MINIMUM	45	15	8	Not Specified	Not Specified
PERSONNEL PER TEAM SUPPORT	Same as Type 2, PLUS: 1 - NIMS Type 2 Veterinarian 1 - NIMS Type 1 Veterinary Assistant 1 - NIMS Type 2 Veterinary Assistant 1 - NIMS Type 1 Animal Behaviorist 4 - NIMS Type 2 Animal Decontamination Specialist 20 - NIMS Type 2 Animal Care and Handling Specialist 1 - Logistics Specialist	Same as Type 3, PLUS: 1 - NIMS Type 2 Veterinarian 1 - NIMS Type 2 Veterinary Assistant 2 - NIMS Type 2 Animal Decontamination Specialist 2 - NIMS Type 2 Animal Care and Handling Specialist 1 - Logistics Specialist	1 - NIMS Type 1 Veterinary Assistant 1 - NIMS Type 1 Animal Care and Handling Specialist 5 - NIMS Type 2 Animal Care and Handling Specialist	Not Specified	1. Add drivers based on the number of vehicles used for transport. Driver is not a NIMS typed position. 2. A driver may be one of the team members, depending on the vehicle and the team member's driving credentials. 3. Logistics Specialist is not a NIMS typed position.
TRANSPORTATION PER TEAM SUPPORT	Same as Type 2	Same as Type 3	1. Vehicles for transporting equipment and supplies, based on mission 2. Passenger vehicles, as appropriate for mission	Not Specified	

FEMA

Resource Typing Definition for Environmental Response/Health and Safety
Animal Emergency Response

NOTES

1. Nationally typed resources represent the minimum criteria for the associated component and capability.
2. Companion animal decontamination teams are challenging to equip, train, and maintain. No two teams have identical equipment. Additional research in animal decontamination, particularly for radiological incidents, can help in identifying a common denominator for team development and deployment. Additional research in animal decontamination, particularly for radiological incidents, can help in identifying equipment and protocols to improve team throughput.
3. See National Alliance of State Animal and Agricultural Emergency Programs (NASAAEP) Animal Decontamination Best Practices for additional information.

REFERENCES

1. FEMA, NIMS 509: Animal Emergency Response Team Leader
2. FEMA, NIMS 509: Veterinarian
3. FEMA, NIMS 509: Veterinary Assistant
4. FEMA, NIMS 509: Animal Behaviorist
5. FEMA, NIMS 509: Animal Decontamination Specialist
6. FEMA, NIMS 509: Animal Care and Handling Specialist
7. FEMA Radiological Emergency Preparedness (REP) Program Manual, FEMA P-1028, 2016
8. Occupational Safety and Health Administration (OSHA) 29 Code of Federal Regulations (CFR) Part 1910.132: Personal Protective Equipment, latest edition adopted
9. National Alliance of State Animal and Agricultural Emergency Programs (NASAAEP) Animal Decontamination Best Practices, 2012

FEMA

Position Qualification for Public Health, Healthcare, and Emergency Medical Services
Animal Emergency Response

VETERINARIAN

RESOURCE CATEGORY	Animal Emergency Response
RESOURCE KIND	Not Specified
OVERALL FUNCTION	
COMPOSITION AND ORDERING SPECIFICATIONS	1. This position can be ordered as a single resource or in conjunction with a NIMS typed team (Veterinary Medical Team).
	2. Discuss logistics for deploying this position, such as security, lodging, transportation, and meals, prior to deployment
	3. This position typically works 12 hours per shift, is self-sustainable for 72 hours, and is deployable for up to 14 days
	4. For Type 1 and Type 2, requestor specifies competency areas necessary based on the animal population the position will serve
	5. For Type 1, requestor specifies specialty areas necessary based on incident requirements

Each type of resource builds on the qualifications of the type below it. For example, Type 1 qualifications include the qualifications in Type 2, plus an increase in capability. Type 1 is the highest qualification level.

COMPONENT	TYPE 1	TYPE 2	NOTES
DESCRIPTION	Same as Type 2, PLUS: Is board certified in a specialty area of veterinary medicine, such as: 1. Anesthesia 2. Dentistry 3. Dermatology 4. Emergency and critical care 5. Internal medicine 6. Radiology 7. Surgery 8. Theriogenology 9. Toxicology	The Type 2 Veterinarian is a general practitioner and: 1. Treats animals in one or more of the following competency areas: a. Companion animals, including pets, service animals, and assistance animals b. Livestock, including food or fiber animals and domesticated equine species c. Wildlife animals, captive wildlife, and zoo animals d. Laboratory animals 2. Investigates cases of animal disease 3. Triages ill or injured animals 4. Performs clinical examinations 5. Diagnoses animal diseases 6. Identifies abnormal conditions in animals 7. Recommends risk mitigation procedures for animal-to-animal and animal-to-human (zoonotic) disease transmission 8. Supervises animal disease control field operations 9. Monitors for the emergence and reemergence of diseases 10. Provides healthcare to animals 11. Advises on animal depopulation 12. Performs euthanasia 13. Oversees vaccination of animals 14. Advises on disease control and prevention 15. Monitors and recommends humane animal care standards 16. Supervises the Veterinary Assistant	Not Specified

FEMA

Position Qualification for Public Health, Healthcare, and Emergency Medical Services
Animal Emergency Response

COMPONENT	TYPE 1	TYPE 2	NOTES
EDUCATION	Same as Type 2, PLUS: Completion of post-graduate education in a specialty area	Doctor of Veterinary Medicine or equivalent degree	Not Specified
TRAINING	Same as Type 2	Completion of the following: 1. IS-100: Introduction to the Incident Command System, ICS-100 2. IS-200: Incident Command System for Single Resources and Initial Action Incidents 3. IS-700: National Incident Management System, An Introduction 4. IS-800: National Response Framework, An Introduction	Not Specified
EXPERIENCE	Same as Type 2, PLUS: Two years of experience in a specialty area	Two years of experience in a clinical setting commensurate with the mission	Not Specified
PHYSICAL/MEDICAL FITNESS	Same as Type 2	1. Performs duties under moderate circumstances characterized by working consecutive 12-hour days under physical and emotional stress for sustained periods of time 2. Is able to work while wearing appropriate Personal Protective Equipment (PPE) 3. Keeps immunizations up to date and commensurate with mission	PPE is mission specific and may vary by work environment; it includes protective footwear, protective clothing for skin exposure, eye and ear protection, respirators, gloves, and masks.
CURRENCY	Same as Type 2	Routinely provides direct patient care in the competency area specified	Not Specified
PROFESSIONAL AND TECHNICAL LICENSES AND CERTIFICATIONS	Same as Type 2, PLUS: Board certification in a specialty area	State, District of Columbia, or U.S. tribal- or territory-granted active status of legal authority to function as a Veterinarian	For some responses, U.S. Department of Agriculture (USDA) accreditation may be necessary.

FEMA

Position Qualification for Public Health, Healthcare, and Emergency Medical Services
Animal Emergency Response

NOTES
Nationally typed resources represent the minimum criteria for the associated category.

REFERENCES
1. FEMA, NIMS 508: Veterinary Medical Team
2. FEMA, NIMS 509: Veterinary Assistant
3. FEMA, NIMS Guideline for the National Qualification System, November 2017

FEMA

Position Qualification for Public Health, Healthcare, and Emergency Medical Services
Animal Emergency Response

VETERINARY ASSISTANT

RESOURCE CATEGORY	Animal Emergency Response
RESOURCE KIND	Not Specified
OVERALL FUNCTION	
COMPOSITION AND ORDERING SPECIFICATIONS	1. This position can be ordered as a single resource or in conjunction with a NIMS typed team (Veterinary Medical Team). 2. Discuss logistics for deploying this position, such as security, lodging, transportation, and meals, prior to deployment 3. This position typically works 12 hours per shift, is self-sustainable for 72 hours, and is deployable for up to 14 days 4. Requestor specifies competency areas necessary based on the animal population the position will serve

Each type of resource builds on the qualifications of the type below it. For example, Type 1 qualifications include the qualifications in Type 2, plus an increase in capability. Type 1 is the highest qualification level.

COMPONENT	TYPE 1	TYPE 2	NOTES
DESCRIPTION	Same as Type 2, PLUS: The Type 1 Veterinary Assistant has a Veterinary Technician or equivalent degree, certificate, or formal training and: 1. Helps administer anesthesia and diagnostic imaging under the supervision of a Veterinarian 2. Performs other duties commensurate with state board regulations for the position	The Type 2 Veterinary Assistant assists a Veterinarian; specifically, this position: 1. Performs veterinary support duties under the direct supervision of a Veterinarian and administers medical care to ill or injured animals in one or more of the following competency areas: a. Companion animals, including pets, service animals, and assistance animals b. Livestock, including food or fiber animals and domesticated equine species c. Wildlife animals, captive wildlife, and zoo animals d. Laboratory animals 2. Provides proper handling and restraint of animals 3. Provides basic medical care to animals 4. Maintains medical records 5. Performs laboratory tests 6. Assists with other veterinary support duties under the direct supervision of a Veterinarian	Not Specified
EDUCATION	Completion of a Veterinary Technician or equivalent degree, certificate, or formal training	Not Applicable	Students in accredited Veterinary, Veterinary Technology, Veterinary Assistant, or equivalent programs may deploy as Type 2 Veterinary Assistants if they meet the training requirements below.

FEMA

Position Qualification for Public Health, Healthcare, and Emergency Medical Services
Animal Emergency Response

COMPONENT	TYPE 1	TYPE 2	NOTES
TRAINING	Same as Type 2	Completion of the following: 1. IS-100: Introduction to the Incident Command System, ICS-100 2. IS-200: Incident Command System for Single Resources and Initial Action Incidents 3. IS-700: National Incident Management System, An Introduction 4. IS-800: National Response Framework, An Introduction	Not Specified
EXPERIENCE	Same as Type 2	Two years of experience in a practice setting commensurate with the mission	Not Specified
PHYSICAL/MEDICAL FITNESS	Same as Type 2	1. Performs duties under moderate circumstances characterized by working consecutive 12-hour days under physical and emotional stress for sustained periods of time 2. Is able to work while wearing appropriate Personal Protective Equipment (PPE) 3. Keeps immunizations up to date and commensurate with mission	PPE is mission specific and may vary by work environment; it includes protective footwear, protective clothing for skin exposure, eye and ear protection, respirators, gloves, and masks.
CURRENCY	Same as Type 2	Routinely provides direct patient care in the competency area specified	Not Specified
PROFESSIONAL AND TECHNICAL LICENSES AND CERTIFICATIONS	State, District of Columbia, or U.S. tribal- or territory-granted active status of legal authority to function as a Veterinary Technician or equivalent, if applicable	Not Specified	Not Specified

FEMA

Position Qualification for Public Health, Healthcare, and Emergency Medical Services
Animal Emergency Response

NOTES

Nationally typed resources represent the minimum criteria for the associated component and capability.

REFERENCES

1. FEMA, NIMS 508: Veterinary Medical Team
2. FEMA, NIMS 509: Veterinarian
3. FEMA, NIMS Guideline for the National Qualification System, November 2017

FEMA

Resource Typing Definition for Public Health, Healthcare, and Emergency Medical Services
Animal Emergency Response

VETERINARY MEDICAL TEAM

DESCRIPTION	The Veterinary Medical Team provides care to animals in a field setting or an existing veterinary medical facility, such as a veterinary hospital, animal shelter, mobile medical unit, or portable temporary hospital setting
RESOURCE CATEGORY	Animal Emergency Response
RESOURCE KIND	Team
OVERALL FUNCTION	This team provides medical care for animals in one or more of the following populations: 1. Companion animals, including pets, service animals, and assistance animals 2. Livestock, including food or fiber animals and domesticated equine species 3. Wildlife, captive wildlife, or zoo animals 4. Laboratory animals
COMPOSITION AND ORDERING SPECIFICATIONS	1. Discuss logistics for deploying this team, such as security, lodging, transportation, and meals, prior to deployment 2. This team typically works 12 hours per shift, is self-sustainable for 72 hours, and is deployable for up to 14 days 3. Requestor specifies competency areas necessary based on the animal population this team will serve, including specific species 4. Requestor specifies approximate number of animals requiring treatment, so provider can estimate appropriate resources and staffing 5. Requestor specifies treatment location 6. Requestor supplements clinical staff by ordering a Healthcare Logistics Coordination and Support Team if logistics, equipment, supplies, and Information Technology/communications support are necessary 7. Individuals responding with this team possess the clinical knowledge and skills necessary to function in the clinical area the requestor specifies 8. Requestor agrees to accept all credentials of the provider's team members, allowing the supplied team to function in existing facilities 9. Requestor orders support specialists such as Pharmacists and Pharmacy Technicians separately 10. Requestor orders additional specialists as necessary—specialists in anesthesia, dentistry, dermatology, emergency and critical care, internal medicine, radiology, surgery, theriogenology, toxicology, and so on 11. Type 2, Type 3, and Type 4 teams provide acute and primary care to stabilize critical patients so the patients can evacuate to intact veterinary facilities outside the disaster area 12. The Type 1 team supports an affected area when animals cannot evacuate to fixed facilities outside the area because of distance or transportation issues

Each type of resource builds on the qualifications of the type below it. For example, Type 1 qualifications include the qualifications in Type 2, plus an increase in capability. Type 1 is the highest qualification level.

FEMA

Resource Typing Definition for Public Health, Healthcare, and Emergency Medical Services
Animal Emergency Response

COMPONENT	TYPE 1	TYPE 2	TYPE 3	TYPE 4	NOTES
CAPABILITIES PER TEAM MEDICAL	Same as Type 2, PLUS: 1. Tertiary care 2. Hospitalization 3. Post-decontamination treatment and monitoring	Same as Type 3, PLUS: 1. Surgery 2. Radiography	Same as Type 4, PLUS: Minor surgery	This team provides: 1. Acute and primary care support 2. Health screening	1. Team members have medical and handling expertise specific to the competency area and animal population they will serve. 2. Depending on incident conditions and species served, Type 4 teams can examine and treat 10 to 50 working animals or 75 to 100 noncritical animal patients per day, on average. 3. Requestor provides equipment for surgery and radiography, unless the provider otherwise specifies.
CAPABILITIES PER TEAM TREATMENT LOCATION	Same as Type 2	Same as Type 3	This team provides care in portable/mobile temporary veterinary medical settings or as field response	This team provides care at an existing facility, such as a veterinary hospital or a shelter, or as field response	Equipment needs depend on competency area.
EQUIPMENT PER TEAM MEDICAL	Same as Type 2	Same as Type 3, PLUS: Major surgical packs	Same as Type 4, PLUS: 1. Minor surgical packs 2. Anesthetic drugs and delivery system	Basic primary and acute care veterinary medical supplies and equipment	1. Requestor provides equipment for surgery and radiography, unless the provider otherwise specifies. 2. Teams may also need species-specific handling equipment, such as halters, lead ropes, lariat ropes, cattle panels, squeeze chutes, leather gloves, muzzles, leashes, and traps.
EQUIPMENT PER TEAM MEMBER COMMUNICATIONS	Same as Type 2	Same as Type 3	Same as Type 4	1. Two-way portable radio 2. Cell phone	Not Specified
EQUIPMENT PER TEAM MEMBER PERSONAL PROTECTIVE EQUIPMENT	Same as Type 2	Same as Type 3	Same as Type 4	PPE is mission specific and may vary by work environment; it includes: 1. Protective footwear 2. Protective clothing for skin exposure 3. Eye and ear protection 4. Respirators 5. Gloves 6. Masks	The following regulation addresses PPE: Occupational Safety and Health Administration (OSHA) 29 Code of Federal Regulations (CFR) Part 1910.132: Personal Protective Equipment.

FEMA

Resource Typing Definition for Public Health, Healthcare, and Emergency Medical Services
Animal Emergency Response

COMPONENT	TYPE 1	TYPE 2	TYPE 3	TYPE 4	NOTES
PERSONNEL PER TEAM MANAGEMENT AND OVERSIGHT	Same as Type 2	Same as Type 3	1 - National Incident Management System (NIMS) Type 1 Animal Emergency Response Team Leader	Not Specified	For Type 4 teams, any team member having completed ICS 300 may function as team leader, or the existing chain of command provides management and oversight at the requestor's discretion.
PERSONNEL PER TEAM MINIMUM	13	10	7	3	Not Specified
PERSONNEL PER TEAM SUPPORT	Same as Type 2, PLUS: 1 - NIMS Type 2 Veterinarian 2 -NIMS Type 2 Veterinary Assistant	Same as Type 3, PLUS: 1 - NIMS Type 2 Veterinarian 2 - NIMS Type 2 Veterinary Assistant	Same as Type 4, PLUS: 1 - NIMS Type 2 Veterinarian 2 - NIMS Type 2 Veterinary Assistant	1 - NIMS Type 2 Veterinarian 2 - Type 2 Veterinary Assistant	1. Requestor may substitute a Type 1 Veterinarian certified in a specific specialty area based on incident requirements. 2. Requestor may substitute Type 1 Veterinary Assistants for Type 2 Veterinary Assistants based on incident requirements.

FEMA

Resource Typing Definition for Public Health, Healthcare, and Emergency Medical Services
Animal Emergency Response

NOTES

Nationally typed resources represent the minimum criteria for the associated component and capability.

REFERENCES

1. FEMA, NIMS 508: Healthcare Logistics Coordination and Support Team
2. FEMA, NIMS 509: Animal Emergency Response Team Leader
3. FEMA, NIMS 509: Pharmacist
4. FEMA, NIMS 509: Pharmacy Technician
5. FEMA, NIMS 509: Veterinarian
6. FEMA, NIMS 509: Veterinary Assistant
7. Occupational Safety and Health Administration (OSHA) 29 Code of Federal Regulations (CFR) Part 1910.132: Personal Protective Equipment, latest edition adopted
8. National Association of State Public Health Veterinarians (NASPHV), Compendia pages containing Veterinary Standards and Infection Control Measures, latest edition adopted
9. National Wildfire Coordinating Group (NWCG), National Incident Management System Wildland Fire Qualification System Guide, PMS 310-1, Physical Fitness Levels, October 2016

APPENDIX B: ASPCA COMMUNITY PREPAREDNESS CHECKLIST

Animal Emergency Preparedness

Self-Assessment Checklist

This checklist is intended to help identify where a community is well resourced and where additional support may be needed.

The checklist will be most effective in identifying strengths and shortfalls when completed by the agency/organization that has primary jurisdictional responsibilities for animals in emergencies and is familiar with the community's Emergency Operations Plan (EOP).

This assessment should be viewed not only as an analysis tool but also as a blueprint for improvements in your animal response plans. If you're in the initial phase of planning, we suggest you first focus on high-priority items marked with this symbol k.

We also encourage you to meet with appropriate partners in your community to formulate next steps to address gaps identified by the checklist.

The fillable, seven-page checklist can be completed, saved, and printed if desired. Use the CLEAR button on the last page if you wish to clear the form and start over. This form works best using Adobe Acrobat Reader.

Acronyms confusing you?

Find most commonly used emergency/disaster acronyms here.

Name of jurisdiction being assessed:	
Jurisdiction definition (city, town, county):	
Date assessment completed:	
Name and title of person(s) completing form:	

Criteria questions $\vdash$ Denotes high priority item	Present	Comments, notes on current status	Proposed next steps, actions
Essential infrastructure			

$\vdash$ **Does an animal plan for emergencies/disasters exist in one or more of the following ways?**
 	
a. As a free-standing Emergency Support Function
 	
b. As part of an ESF, such as
  • Mass Care, Housing and Human Services
  • Public Health & Medical Services
  • Agriculture & Natural Resources
 	
c. As an annex or addendum to the Emergency Operations Plan
 	
Disaster planning tools
	☐		

$\vdash$ **Is a Community Animal Response Team (CART):** **a.** Established **b.** Active **c.** A recognized member of emergency management system	a. ☐ b. ☐ c. ☐		
$\vdash$ **Are the animal response community, Authority Having Jurisdiction (AHJ) and CART recognized members of the Emergency Management community?** To qualify as recognized, must be listed as primary or secondary contributors in the animal annex of the Emergency Operations Plan (EOP).	☐		

Criteria questions K Denotes high priority item	Present	Comments, notes on current status	Proposed next steps, actions
Organization & leadership			
Is an Animal Issues Committee comprised of planning partners (e.g. Fire, Law Enforcement, Animal Control, animal welfare groups, American Red Cross, Emergency Management, etc.) established?	☐		
Does a mutual aid or inter-local agreement with neighboring counties describe types of assistance that might be needed and financial responsibilities when requested?	☐		
Is a Memorandum of Understanding (MOU) or Mutual Aid Agreement (MAA) with community or national animal welfare group established? Sample MOU _____	☐		
Written plan elements	**Date of last plan review:**		
Have the most common emergency/disaster type(s) for the community and potential impact to animals been identified?	☐		
Is the plan broad enough to remain current for at least 3 years? Sample plan _____	☐		

Criteria questions k Denotes high priority item	Present	Comments, notes on current status	Proposed next steps, actions
Written plan elements *(continued)*			
Have specific protocols and organizational chart been identified, plus where animal response fits into county org chart?	☐		
k **Are animal shelter set- up instructions and shelter operations plan in place?** Sheltering tools _____	☐		
Have number and species of animals who may require assistance during emergency/disaster been identified? Pet ownership calculator _____	☐		
Have animal species eligible for evacuation in the community been named?	☐		
Have all potential sources of impacted animals (shelters, boarding, groomers, research facilities, breeders, veterinary clinics, zoos, etc.) been identified?	☐		
Have those who can provide assistance for species requiring special assistance (e.g. large, injured, aggressive animals) been identified and confirmed?	☐		

Criteria questions ʞ Denotes high priority item	Present	Comments, notes on current status	Proposed next steps, actions
Written plan elements *(continued)*			
Have those who can provide assistance or special handling for animals who cannot be evacuated or are abandoned been identified and confirmed?	☐		
Has the service area (legal jurisdiction) been defined, and those who will need animal-related assistance been identified?	☐		
Have agencies/groups been identified to provide evacuations for people with their pets?	☐		
Equipment & systems			
Has a cache been established of necessary equipment and supplies to provide emergency sheltering for at least 50 small animals?	☐		
Has a cache been established of necessary equipment and supplies to provide emergency sheltering for at least 50 large animals?	☐		

Criteria questions ⎡ Denotes high priority item	Present	Comments, notes on current status	Proposed next steps, actions
Rescue, sheltering & reunification			
Is at least one emergency shelter a pet-friendly facility (allows people/ pet cohabitation) or located near available pet housing?	☐		
Has an agency/group responsible for small animal search and rescue been identified?	☐		
Has an agency/group responsible for large animal search and rescue been identified?	☐		
Has an agency/group responsible for providing sheltering support for small animals been identified?	☐		
Has an agency/group responsible for providing sheltering support for large animals been identified?	☐		
Have protocols for critical animal response functions been identified?	☐		
Is a reunification plan in place to ensure people and animals are reunited? Reunification tools _____	☐		

Criteria questions к Denotes high priority item	Present	Comments, notes on current status	Proposed next steps, actions
Personnel, volunteers, training			
For crisis communications, has an individual/branch been designated responsible for streamlining communication and ensuring effectiveness of planning and response?	☐		
Has an agency public information officer (PIO) been identified for agency communication and effective messaging?	☐		
Have personnel been assigned to update and maintain the written plan?	☐		
Is the contact list for staff, volunteers and collaborating agencies current?	☐		
Have volunteers been recruited and trained to assist with animal management? Volunteer management tools	☐		
к **For shelter staffing capacity, are one or both of the following present?** a. Capabilities and resources to provide shelter and care for at least 72 hours b. Plan in place, such as an MOU, to bring in sufficient staff to cover shelter for at least 72 hours	☐		

Criteria questions k Denotes high priority item	Present	Comments, notes on current status	Proposed next steps, actions
Personnel, volunteers, training *(continued)*			
Have small/large animal response teams been created in compliance with established system?	☐		
Is there a program in place for recruiting, training and equipping volunteers to ensure adequate animal response capabilities?	☐		

Additional comments

print

CleaR

Version 1 - 7/2017

APPENDIX C: EQUIPMENT FOUND IN CAMET[1] TRAILER

1.5 × 11 Poly envelopes for cages
200 Animal intake registration forms
250 Tab band collars for animal identification
1250 Colored wrist bands for owner identification
1250 Registration log books
5 Polaroid cameras and 10 film cassettes
1000 Cable ties for envelopes on animal cages
Flashlight
2 Plastic 100 × 16 or 20 ft wide to line walls
3 Corrugated mat 100 ft rolls for center aisle
1 Painter's tape for plastic on walls
6 Disposable rubber gloves—pairs
200 Buckets
5 Scrub brushes
5 Pooper scoopers
5 Trash containers
2 Garbage bags
50 Boxes of plastic bags for feces pick up
5 Hand washing stations
1 Quaternary disinfectant—bottle
1 Bleach—bottle
1 Spray bottle for cleaning cages/crates
2 Litter boxes
5 Litter scoops
5 Cat litters—pounds
100 Orange cones for inside walking area
1 Cage/crate—large 48″ × 30″ × 36″
45 Cages/crates—medium 30″ × 19″ × 22″
15 Bowls
50 Spoons/measuring cups to scoop food
2 Can openers
2 Hoses 1″ × 50″
1 Mop
1 Mop bucket with wringer
1 Shop vacuum
1 Pressure washer
1 Jack stands (pair)
2 Animal control poles
1 Handling gloves
2 5 kW Generators
1 Gas can for generator
1 Telescoping work light
2 First-aid kits (for volunteers and pets)

[1] http://sartusa.org/wp-content/uploads/2013/11/sart-camet-083007.pdf.

APPENDIX D: LARGE ANIMAL EQUIPMENT CACHE[2]

Straps

- At least two rescue straps, 2-ply web, 4″ × 18 ft long with large (10–12 in) flat loops on each end. Suggest having a third strap about 12 ft long for smaller animals

Webbing

- 1″ web, several lengths 20′–25′ long, also several shorter lengths 6, 10, and 15 ft

Rescue Glide

- With slip sheet (preferably two slip sheets)
- With two lengths of 2″ web (single ply) 35′ ft long each

Head Protectors

- Shanks veterinary supply
- Saddle blanket such as a Toklat "cool back"

Generator

- Portable 1000 or 2000 W

Lights

- 2 × 300 W, extension cords (25′ and 50′)

Rope Systems

- Preconstructed 4:1, 3:1, and 2:1 systems made with 5/8″ rope, heavy-duty carabiners, gathering plates, and prusiks (Tandem).
- Anchor/haul rope (200–300 ft) (preferably two ropes).
- Several lengths of softer, flexible rope (25–30 ft).
- Two or three lengths of soft kernmantle rope (25 ft long) to be used for long lead lines
- Six lengths of web (2″ wide, 8–12 ft long with flat loops on each end, "cargo straps") are used to wrap anchors to connect rope systems, also used for trailer upright and lower.
- Additional pullies (4), carabiners (10), and tandem prusik sets (6).
- Throw line.

Telescopic boat hook

- Pike pole or snake tongs or "grabbers"

Miscellaneous

- Earplugs (stuff nylon stockings with cotton balls), blind fold (fly mask with towel insert), hobbles (preferably with quick release), various halters, vet lubricant, stethoscope for taking vitals, medical gloves, first aid bandages, vet wrap, blankets, and saddle blankets

Medical bag

- Oxygen, stethoscope, bandages, tape, vet wrap, BP cuff. Include medical supplies for the animal and the rescuers.

Equipment for trailer manipulation

- Four 2″ webbing, 25 ft long with flat loops on each end
- Rescue 8 heavy-duty friction devices or a brake rack

APPENDIX E: SAMPLE ASSESSMENT FORM[3]

Initial damage assessment form (animal focused)		
Date of assessment:	**Requesting agency & contact info:**	**Assessor & contact info:**
GPS or physical address:	**County/Parish**	***Emergency needs*** ☐No ☐ Yes
Owner/business information (Last, first, middle or name of business; phone # and/or email)		
1. Disaster event (name or type, and date)	**2. Accessibility:** Air Boat Road	

3. Animal damage information

	Roaming	Injured	Missing	Carcasses	Good	Sustainable	Poor
Pets							
Cattle							
Dairy							
Equine							
Sheep/goats							
Captive wildlife							
Exotics/pocket pets							
Comm. poultry							
Overall animal health	(report by species; Good – Sustainable – Poor)						
Water source	Fresh water Flood water Salt water None						
Animal food	1 month supply 1 week supply 3 day supply None						

4. Wildlife interactions

Did you experience any wildlife interactions?

☐No   ☐Yes   If yes, please list species and location and continue below:

_____   _____

If yes, was the wildlife activity normal or abnormal for the species?

☐Normal   ☐Abnormal   ☐Not sure

(Continued)

[3] NASAAEP BPWG 2015.

5. Crop losses (excluding timber)

Total estimated acres destroyed in county/parish :_____

6. Infrastructure losses in county/parish

	# Destroyed	# Open for business
Private veterinary clinics		
Pet shelters (include city locations under #8)		
Animal feed & supply stores		
Livestock sale markets		
Boarding facilities and fairgrounds		
Commercial pet food & supplies		
Fencing (report by # of farms observed)		

7. Immediate needs

	# Pets	# Livestock
Veterinary assistance		
Food		
Livestock containment equipment		
Shelter		
Livestock identification		

8. Assessor recommendations:

☐ *Reassessment in _____ days.*

☐ *Send in Task Force of _____ responders with specialized experience in:*

_____.

☐ *No follow-up needed*

☐ *Pet shelter locations*

(cities):_____.

APPENDIX F: REQUEST FOR ASSISTANCE FORM

Date: [   ]   Submitted by: [   ]

Requesting agency: [   ]

Agency address: [   ]

Agency contact: [   ]   Phone: [   ]

Fax: [   ]   Cell: [   ]

Email: [   ]

Type of assistance request: Choose from the pulldown menu.

[Choose an item.]

Assistance request summary: Please summarize the incident you are requesting assistance for.

[   ]

Location of incident: [   ]

Location of staging: (If different from Incident) [   ]

Type of incident/disaster: Choose from the pulldown menu.

[Choose an item.]

Species affected: Please list all [   ]

Number of animals affected: (If multiple species affected, please include numbers for each)

[   ]

If event is a natural disaster: **Please summarize hazards (contamination; aggressive animals; fire; power inaccessible, etc.), area population and houses affected.**

[   ]

Other agencies contacted for assistance: (Please list any other agencies you've contacted)

[   ]

Community resources available: (What resources are available in the affected community?)

[   ]

Resources requested: (Please specify the types of resources you are requesting from the ASPCA)

[   ]

APPENDIX G: EMERGENCY ANIMAL FACILITY CHECKLIST

Plumbing	Comments
Number and placement of water outlets	
Number and placement of hot water outlets	
Drinking fountains	
Number and location of floor drains	
Fire sprinklers/working order	
Outside water outlets	
HVAC	
Are heating/AC units working properly/type?	
Number/locations of thermostats	
Structural	
Approximate square footage—usable space	
Working windows and are they intact?	
Roof condition/any noticeable water damage from inside or outside	
Floor material and condition	
Number of doors and condition	
Ventilation/airflow?	
Security	
Is there an existing system that can be activated (e.g., ADT)?	
How many entry points to the building needs to be secured by the system?	
Are all dock doors in working order?	
Do the dock plates work for load/unload purposes?	
Are all entry doors working properly (can they lock/unlock easily)?	
Is the building exterior lighting working?	
Electric	
Is there adequate electric service for our needs (amperage)/condition?	
Number and placement of electric runs and capacity	
Adequate lighting?	
Outside electric/capacity	
IT	
Cell coverage	
Landline/DSL	
Restrooms	
Number of toilets/sinks/condition?	
Hot water in restrooms?	
Office	

Plumbing	Comments
Sufficient electric outlets to support computers/printers	
Separate entrance?	
Storage	
Approximate square footage	
Securable space?	
Safety	
Number and location of fire extinguishers	
Location of closest fire hydrant	
Building address and entrance visible from street for emergency service response	
Location of nearest EMS unit and hospital	
Fuel tanks/pumps/other hazardous material?	
Other Operational Considerations	
Snow removal: roof, sidewalks, driveways	
Smoking area	
Parking area for trailers	
Common area for lunch/rehab	
Time frame restrictions	
Transport vehicles accessible—room to maneuver (load and offload)	
Ability to park trailers/office trailer	
Parking for responders	

APPENDIX H: SHELTERING INTAKE FORM

Animal intake form

Incident: _____

Intake personnel name: _____ Title: _____

Date: _____ Case # _____ Animal ID # _____

Animal transport # _____ Agency or team: _____

Animal stats

Name	Species	Breed	Color/markings	Gender	Known ID
	☐ Dog ☐ Cat ☐ Other _____			☐ Female ☐ Male Altered ☐ Yes ☐ No	☐ Collar ☐ ID Tag ☐ License: _____ ☐ Rabies:_____ ☐ Microchip:_____ ☐ Tattoo:_____

Initial Evaluation

Behavior: ☐ Friendly ☐ Shy/cautious/ fearful ☐ Aggressive ☐ Biter/bite hold

Animal health status: ☐ Emergency medical ☐ Medical care advised ☐ Stable ☐ Pregnant ☐ Deceased

Medical exam: Date: _____ Veterinarian: _____ ☐ Examined ☐ Treatment sheet filed

Final evaluation

Disposition: ☐ Returned to owner ☐ Deceased ☐ Euthanized

☐ Adopted
Name: _____
Address: _____
Phone: ( ) _____ ( ) _____

☐ Transferred
Organization: _____
Address: _____
Contact: _____
Phone: ( ) _____ ( ) _____

Final Behavioral Evaluation

Behavior: ☐ Friendly ☐ Shy/cautious/ fearful ☐ Aggressive ☐ Biter/bite hold

Exit personnel/evaluator name: _____ **Title:**_____

APPENDIX I: REQUEST FOR RESCUE FORM

Request for rescue form

Incident: _____

Date: _____   Time: _____ am/pm   Animal ID # _____

Reason for rescue: ☐ Owner request ☐ Agent request ☐ EM/IC ☐ ACO

☐ ASAR ☐ Other: _____

Team assigned: _____

_____   _____

Address/current location of animal   Additional information for rescue

_____   _____

Owner(s) name   Address (city, state, zip)

_____   (   )_____   (   )_____

Owner email address   Phone   Cell phone

_____   _____ (   )_____ (   )_____

Person requesting rescue   Relationship   Phone   Cell phone

_____   (   )_____

Veterinarian or hospital/office name   Phone

Animal information

Species	Breed	Color/markings	Behavior
☐ Dog ☐ Cat ☐ Other: _____			☐ Friendly ☐ Shy/cautious/fearful ☐ Aggressive ☐ Biter/hold

Please initial where consent will be given:

____ Does the person requesting rescue have the owner(s) permission to authorize necessary care? ☐ Y ☐ N

____ Is authorization provided for in field medical care? ☐ Y ☐ N Do not perform: _____

_____

____ Is key available? ☐ Y ☐ N Key location: _____

____ Is keyless entry authorized? ☐ Y ☐ N Means of approved entry? _____

_____   _____   _____

Signature   Print   Date

Status: ☐ Rescued ☐ Unable to capture ☐ No sign of animal ☐ Access denied ☐ Found deceased

---

Animal health: ☐ Emergency care ☐ Needs medical care ☐ Stable ☐ Pregnant

---

Outcome: ☐ Returned to owner(s)

Transported: ☐ Emergency shelter ☐ Veterinarian ☐ Foster ☐ Other

Location: _____

APPENDIX J: ASPCA TRANSPORT MANIFEST FORM

Case #: _____   Date: _____

Truck #: _____   Drivers

Name: _____

Origin site:	Destination site:
Departure time:	Arrival time:

	Animal ID #	Breed	Color	# of Young	Comments
1					
2					
3					
4					
5					
6					
7					
8					
9					
10					
11					
12					
13					
14					
15					
16					
17					
18					
19					
20					
21					
22					
23					
24					
25					

Driver signature: -

_____

Load Master name: _____ Signature: _____

APPENDIX K: ASPCA DAILY OBSERVATION FORM

Date																		
Time	A M	P M	A M	P M	A M	P M	A M	P M	A M	P M	A M	P M	A M	P M	A M	P M	A M	P M
Initials																		
Appetite																		
Good																		
Average																		
Nibbling																		
Not eating																		
Stools																		
Normal																		
Loose																		
Diarrhea																		
Bloody																		
None																		
Urine																		
Normal																		
Excessive																		
Strong odor																		
Bloody																		
Straining																		
None																		
Vomiting																		
Food																		
Bile																		
Hairball																		
Blood																		
COUGHING																		
Sneezing																		
Clear																		
Yellow																		
Green																		
Blood																		
Eyes																		
Clear disch.																		
Pus/mucus																		
Red/irritated																		
Swollen																		
Behavior																		
Friendly greeting																		
Shy but greets																		
Fearful, no greeting																		
Growl/snarl/snap/ bite																		
Able to remove for cleaning																		
Unable to remove																		

Comments:

APPENDIX L: ASPCA PROTOCOL FOR DONNING AND DOFFING PPE IN A HIGH-RISK ENVIRONMENT

Supplies

Tyvek suit—sizes M, L, XL, 2XL, and 3XL
Plastic bags
Shoe covers
N-95 respirator
Nitrile gloves—sizes S, M, L, and XL
Safety goggles or face shield
Duct tape, yellow (for name and "Decon" label)
Duct tape (tape Tyvek sleeves to gloves)
Sharpie
Rubber boots
Rain suit (jacket/pants/hood)—sizes M, L, XL, 2XL, and 3XL
Heavy-duty outer rubber gloves

At the beginning of decon shift

1. Pick up the right size of Tyvek suit and a plastic bag.
2. Go to the bathroom and take off personal clothes that you want to wear at home at the end of the day.
3. Remove all rings, necklaces, jewelry, piercings, and electronics. Place a Band-Aid over piercings.
4. Place all clothes and personal items except for shoes and socks into the plastic bag and put on Tyvek suit.
5. Put socks and shoes back on.
6. Return to where you picked up your Tyvek suit with your plastic bag and wait for further instructions.

Donning PPE

1. Using yellow duct tape, put your first name on the front and "Decon" label on the front and back of Tyvek.
2. Put on shoe covers.
3. Put on N-95 respirator and check for fit.
4. Put on face shield or safety goggles.
5. Put on inner and outer nitrile gloves.
6. Tuck each Tyvek sleeve in between the two nitrile gloves.
7. Duct tape the outer nitrile gloves to Tyvek suit.
8. When directed, enter Decon area.
9. Put rubber boots on.
10. Put on bottom of rain suit.
11. Put on top of rain suit and close.
12. Put outer rubber gloves on with the rain suit over the glove.
13. Working with a partner, tape ankles and wrists as shown.
14. Assist partner with pulling hood on.

Doffing PPE

1. Wait to be sprayed off with fresh water before proceeding to saw horses.
2. Remove all tape and put in trash can.
3. Remove outer rubber gloves.
4. Put gloves over saw horse.
5. Take off rain suit jacket.
6. Hang rain jacket over saw horse.
7. Take off rain suit pants.
8. Hang pants over saw horse.
9. Head toward exit.
10. Take off rubber boots.
11. Take off shoe covers and place in trash can.
12. Take off nitrile gloves and place in trash can.
13. Take off face shield or safety goggles and hold onto them.

14. Take off N-95 respirator and place in trash can.
15. Step outside of the tarped area.
16. Leave face shield or safety goggles on table.
17. Keep Tyvek suit and socks/shoes on to take a break or eat lunch.

At the end of shift

1. Take off Tyvek suit after removing all other PPE.
2. Put personal clothes on.
3. Place used Tyvek in plastic bag.
4. Return plastic bag containing used Tyvek to Decon area for disposal.

APPENDIX M: ASPCA CLEANING PROTOCOL FOR EQUIPMENT

Exposed to Hazardous Materials Equipment

4 625 gal water tanks

2 Virkon tanks

2 Rinse tanks

55 lb of Virkon per tank (10 5.5 lb tubs)

2 Sump pumps plus dedicated lines (two hose lengths per sump pump—one black sump pump line and one green sump pump line)

1 Fresh water line (two black hose lengths, labeled "water")

3 Long handled hard bristle scrub brushes

1 Fan, large

1 Torpedo heater

1 30 gal oval tub (for storing small supplies and to help tip tanks for emptying)

1 Puppy pee pads (package) or five rolls of paper towels

2 Timers (plus one backup timer)

2 Saw horses (approximately 4' high to hang rain suits and heavy-duty outer rubber gloves on to dry)

3 Rolls of duct tape red (labeling biohazardous trash bags for gross decon and contaminated PPE)

10 Rolls of duct tape gray for PPE

2 44 gal Trash cans (one for gross decon and one for PPE disposal)

1 15 gal Trash can (PPE disposal)

1 box trash bags for each trash can

4 Sharpies (two for outside decon area and two for inside decon area)

1 Pliers, diagonal 6"

Personal protective equipment (PPE) for each team member (rain suit, Tyvek suit, nitrile gloves, safety goggles/face shield, shoe covers/rubber boots, heavy-duty outer rubber gloves, N-95 respirator, and yellow duct tape for name and labeling)

1 Radio

Personnel—Total 7 for Maximum Efficiency

1—Bring in contaminated crates/gross decontamination

2—Virkon scrubbers

3—Rinse tank transfer

4—Identification tag removal/transfer to drying area

5—Stack dry crates/count/transfer out clean crate door

6—Runner (handling sump pumps, refilling tanks with water, taping/retaping wrists, handling communication, and taking care of team needs)

Crates/Tank

30 Small crates will fit in one tank at a time.

Note: 15–20 small crates are a more realistic number to ensure that all crates are adequately scrubbed within the 10 min Virkon soak time frame.

9 Medium crates will fit in one tank at a time.

Crate Cleaning Process

1. Contaminated crates enter through the garage door and are stacked until a sufficient number are inside on the right, and then, the garage door is closed.
2. All contaminated crates are sprayed with Virkon from pump sprayer. If it is the first set of crates for the shift, they can instead go directly into the Virkon.
3. Sprayed contaminated crates are turned upside down over a garage can, and all manure/feathers/dust is swept out with a long handled hard bristle scrub brush (gross decontamination) and then stacked on the left.
4. After a Virkon tank is filled with contaminated crates, the tank's timer is set for 10 min. The timers are color coded: red tape on timer for red-taped tank and yellow tape on timer for yellow-taped tank.
5. The two scrubbers then start scrubbing all inside surfaces of the contaminated crates using long handled hard bristle scrub brushes.

6. Once the timer goes off, the crates are transferred from the Virkon tank to the rinse tank as quickly as possible while also draining off the majority of excess liquid so that the Virkon tank can be quickly reloaded with the next batch of contaminated crates.

7. Crates are dunked up and down and/or rolled in the rinse water several times, as much water as possible is drained off, and then, the crate is removed.

8. Any identification tags present are cut off the crates, and then crates are transferred to the drying area where they are placed door down with side vents facing the fan and heater.

9. Once a batch of clean crates is dry (approximately 25–30 min), they are stacked and counted.

10. Once a large number of clean crates have been stacked, the runner will radio operations or logistics for assistance to transfer clean crates to the clean crate section of the bay.

Changing Virkon and Rinse Tank Water

A tank takes approximately 30 min to refill with fresh water.

Sump pumps must be checked for debris that will clog and decrease or stop the sump pump from working. Checking and removing debris are essential when the tank is almost finished draining.

Draining Rinse Water Tank

Using two sump pumps, it takes about 30 min to drain a rinse water tank.

The sump pump attached to the green hose will remove approximately 99 % of the water once the sump pump attached to the black hose has stopped working.

Using both sump pumps, drain each into a separate hand washing sink.

Once approximately 85–90 % of the tank has been drained, put the 30 gal black oval tub underneath one edge of the tank to help sump pumps remove the rest of the water.

The fresh water line can be used to help rinse any remaining residue and debris down and further dilute it so that more will be removed from the tank.

Once the sump pumps stop drawing water, unplug and remove the sump pumps.

Clean out as much of the residue that is left as possible before starting to refill using puppy pee pads or paper towels. Zip ties and other plastic or metal pieces can be removed while wearing gloves.

Remove the black oval tub and reposition the tank.

Start refilling with freshwater line.

Draining Virkon Tank

It will take approximately 2 h to drain one Virkon tank using ONE sump pump ONLY (see notes below).

Using ONLY the sump pump attached to the green hose, start draining the Virkon into the farthest hand washing sink.

Once approximately 85–90 % of the tank has been drained, put the 30 gal black oval tub underneath one edge of the tank to help the sump pump remove the rest of the Virkon.

The freshwater line can be used to help rinse any remaining residue and debris down and further dilute it so that more will be removed from the tank.

Once the sump pump stops drawing water, unplug and remove the sump pump.

Clean out as much of the residue that is left as possible before starting to refill using puppy pee pads or paper towels. Zip ties and other plastic or metal pieces can be removed while wearing gloves.

Remove the black oval tub and reposition the tank.

Start refilling with freshwater line.

Notes:

1. The drains in some hand washing sinks are not capable of handling both sump pumps draining a Virkon tank. Foam will bubble up in all the sinks and toilets.

2. Someone must monitor the hand washing sinks and switch the drain hose back and forth between two sinks to keep the suds down.

3. Cold water must run full force in the hand washing sink that the drain hose is currently in to help decrease/prevent suds both in the sink and in the bathroom plumbing.

Index

Note: Page numbers followed by *f* indicate figures.

Printed in the United States
By Bookmasters